Green Energy and Technology

AF779986

For further volumes:
http://www.springer.com/series/8059

Said Al-Hallaj · Kristofer Kiszynski

Hybrid Hydrogen Systems

Stationary and Transportation Applications

Springer

Dr. Said Al-Hallaj
All Cell Technologies LLC
2321 W. 41st Street
Chicago, IL 60609
USA
e-mail: salhallaj@allcelltech.com

Kristofer Kiszynski
4937 N. Christiana Ave.
Chicago, IL 60625
USA
e-mail: kristofer.kiszynski@shawgrp.com

ISSN 1865-3529 e-ISSN 1865-3537

ISBN 978-1-4471-2650-8 ISBN 978-1-84628-467-0 (eBook)

DOI 10.1007/978-1-84628-467-0

Springer London Dordrecht Heidelberg New York

British Library Cataloguing in Publication Data
A catalogue record for this book is available from the British Library

Softcover reprint of the hardcover 1st edition 2011

Cover design: eStudio Calamar, Berlin/Figueres

Printed on acid-free paper

Springer is part of Springer Science+Business Media (www.springer.com)

Preface

This book was inspired by my need for a textbook to teach the Renewable Energy Technology class that I developed at the Illinois Institute of Technology (IIT) and I am now teaching at the University of Illinois at Chicago (UIC). Most textbooks that cover the subjects of renewable energy and hydrogen systems focus on the supply side with little or no mention of the considerations one must make to balance of the system in the areas of energy storage, system control, and optimization. This book is an attempt to cover that gap by providing a textbook to present basic knowledge of the various components of hybrid renewable hydrogen systems as well as a methodology and overview for systems design, control, and optimization.

This book is intended for an audience of senior undergraduate and graduate engineering students as well as other graduate students from colleges of Architecture, Business, Law or Policy. This book is also intended to discuss the various aspects of the much celebrated "Hydrogen Economy" and to cover the basics of renewable energy sources and their technical and economical limitations or barriers. In addition, this book provides an overview of energy storage and conversion technologies such as hydrogen storage, batteries and fuel cells. Finally, case studies on hybrid hydrogen systems for clean energy and clean water applications are presented.

The first chapter provides an overview of worldwide energy consumption, the state of renewable energy, and the potential role renewable energy can play in a sustainable energy future.

Chapter 2 presents an overview of first principles in solar and wind energy. The chapter discusses fundamentals of operation as well technical and economical constraints.

Chapter 3 provides an overview of hydrogen production methods and discusses various alternatives for hydrogen storage. In addition the chapter provides a good summary for the various types of fuel cells and their operating principles.

Chapter 4 describes the operation of the Renewable Hybrid Energy System (RHES) and explains why hybrid generation and storage make economic sense under certain conditions. The chapter also discusses the dynamic behavior of a fuel cell and battery hybrid system and the design of an active controller for such a system.

Chapter 5 provides an overview of hybrid energy systems, such as the PEM (Polymer Electrolyte Membrane) fuel cell/battery hybrid system, as well as guidelines for control of these systems. The controller logic developed is able to respond to three different load scenarios. The controller is also tuned to buffer the fuel cell from load transients. The work in this chapter is based on previous work from the inter-professional project "Solar Hydrogen Project" at IIT, sponsored by the Illinois Department of Commerce, ComEd, BP, and Proton Energy Systems.

Chapter 6 presents a case study that describes the design and implementation of a hybrid system for the elimination of engine idle in airport ground support vehicles. A PEM (Polymer Electrolyte Membrane) fuel cell/lithium-ion battery system is shown to be the most suitable design for this project. Details of the proposed design are discussed as well results and a summary of the successes and limitations of the project along with proposed future work. The project was sponsored by the Chicago Department of Fleet Management (CDFM) with major technical contributions and insights by Mathew Stewart at CDFM and Mohammed Khader from AllCell Technologies.

Chapter 7 discusses the key decisions that factor into the design of a hybrid fuel cell/desalination (HFCD) system to supply a developing region with adequate electrical power and water. The focus of this case study is Caye Caulker, a Carribean island located off the coast of Belize. Caye Caulker has limited fresh water sources and currently uses diesel generators as its sole source of power. The goal of this case study is to replace these diesel generators with an HCFD system that can also provide Caye Caulker with potable water. The work in this case study was part of a project sponsored by the Middle East Desalination center in Oman. Major contributions were made to the chapter material by Greg Albright.

The authors are grateful to Hisham Teymour and Katherine Lazicki for their assistance in editing and formatting the book. In addition, the authors are thankful to Springer's staff and to Claire Protherough, Senior Editorial Assistant for encouragement and follow-ups.

Dr. Said Al-Hallaj

Contents

Chapter 1
The Role of Renewable Energy in a Sustainable Energy Future

Now overwhelming scientific consensus that fossil fuels are causing serious climate change (Science, December 2004)

The climate is changing at an unnerving pace. Glaciers are retreating, ice shelves are fracturing, sea level is rising, permafrost is melting... How can we not cover the biggest geography story of the century? (National Geographic, September 2004 issue, which devoted 74 pages to the signs of climate change)

1.1 Fossil Fuel Based Economy and Climate Change Challenges

Growth of the human population and global economic activity are placing significant strain on the life-carrying capacity of the Earth. Demands for food, clean water, shelter and energy are increasing as is the pollution generated through the consumption of these resources. The increase in demand for these resources is driven not only by population growth, but also from the desire to improve one's standard of living. In general, an improved standard of living requires an increase in the energy used per person. These two factors together lead to compounded growth in the demand for limited resources and in the production of hazardous pollutants.

Almost all of the energy used for transportation and a significant portion of energy used for stationary applications is derived from fossil fuels. While the burning of fossil fuels has resulted in tremendous economic growth, increased productivity and an improved standard of living in some areas of the world over the last century, it is not sustainable. The use of fossil fuels causes environmental degradation and health problems, these resources are finite and a constant supply of fossil fuels to countries that do not have large amounts of their own are dependent on those countries that do.

S. Al-Hallaj and K. Kiszynski, *Hybrid Hydrogen Systems*,
Green Energy and Technology, DOI: 10.1007/978-1-84628-467-0_1,

The reserves of fossil fuels that currently power society will fall short of this demand over the long term. Many alternative renewable fuels are currently far from competitive with fossil fuels in cost and production capacity. Combustion of fossil fuels produces both gaseous and particulate emissions that can negatively affect both the environment and the health of people. In 1995, global emissions totaled 22.19 billion metric tons (BMT) of CO_2, 852 million metric tons (MMT) of CO, 99.27 MMT of (nitrogen oxides) NO_x, and 141.9 MMT of sulfur dioxide (SO_2) [1]. These values are growing each year and this growth will accelerate as countries with developing economies and large populations such as India and China expand.

At 370 ppm, atmospheric CO_2 levels are currently at their highest level in 420,000 years. Carbon dioxide levels today are 18% higher than in 1960 and an estimated 31% higher than they were at the onset of the Industrial Revolution in 1750 [2]. Historical studies show a strong correlation between atmospheric CO_2 levels and air temperature. The effects of this rise in temperature, also known as global warming, are unknown although predicted scenarios suggest serious consequences. Among the associated environmental impacts are; biodiversity loss, sea level rise, increased drought, spread of disease, weather pattern shifts, increased flooding, changes in freshwater supply, and an increase in extreme weather events [3].

Health care costs associated with the treatment of conditions such as asthma and the loss in worker productivity due to poor health is having and will have an increasing negative impact on the world's economies. NO_x is responsible for the formation of ambient ozone which is created when sunlight is exposed to NO_x and hydrocarbons. It can be an especially serious problem in urban areas with high population densities. Ozone is a respiratory tract irritant and can cause shortness of breath, pain when inhaling, exacerbated asthma symptoms, wheezing and cough in children. Ozone also causes airway inflammation and decreased pulmonary function in adults [4]. Exposure to NO_x can also enhance the allergic response to allergens. Particulate pollution contributes to excess mortality and hospitalization for cardiac and respiratory tract disease in adults [5–8]. Other studies have found that exposure to diesel exhaust, a major source of particulate emissions, is

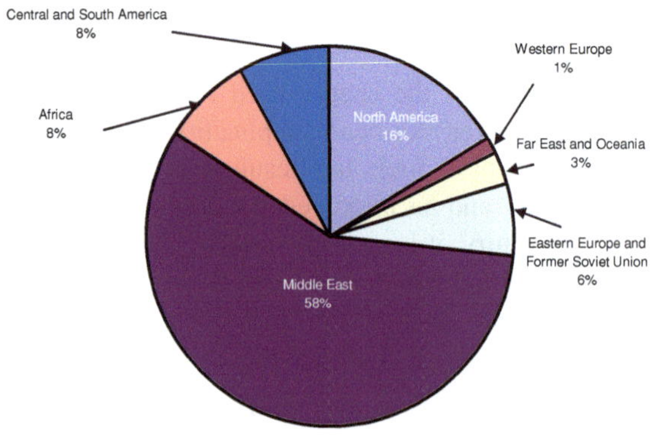

Fig. 1.1 Distribution of worldwide proven oil reserves by region. *Source*: EIA Energy Outlook 2006 Report

associated with increased risk of lung cancer. In children, this type of pollution affects lung function and growth [9]. It also increases the symptoms of bronchitis and other studies found associations between particulate pollution and post-neonatal infant mortality, low birth weight and preterm birth [10–17].

In industrialized nations, an uninterruptible source of energy is critical to economic stability and energy security. As an example, Fig. 1.1 shows the distribution of world's proven oil reserves. As many industrialized countries are highly dependent on other countries for their energy, their economies are dependent on the willingness of other countries to supply it.

Lastly, fossil fuel supplies are limited. Although fossil fuels can be consumed more efficiently through improved energy conversion techniques and other means, these supplies are finite. They are not the ultimate solution to the problem of energy demand and an alternative to meet this demand will need to be developed.

1.2 Review of World Energy Production and Consumption

The world now uses energy at a rate of approximately 1.14×10^{14} kWh/year, equivalent to a continuous power consumption of 13 terawatts (TW). Currently, nearly 81% of the world's electricity is generated from non-renewable resources (41% coal, 17% nuclear, 17% natural gas, and 6% oil) and 19% from hydropower and other renewables (Fig. 1.2). According to the US DOE Energy Information Administration projections in Fig. 1.3 (EIA, Annual Energy Outlook 2006), renewable resources will grow but their overall contribution to world's electricity supply will drop below its current level to 17% in 2030 [18].

The EIA anticipates significant capacity addition to electric generation capacity to meet the growing demand for electricity and to replace old plants. As shown in Fig. 1.4, the majority of these projected additions are expected to come from combined-cycle natural gas plants due to their high efficiency (50% and above).

Fig. 1.2 Share of energy production by chart type (2003)

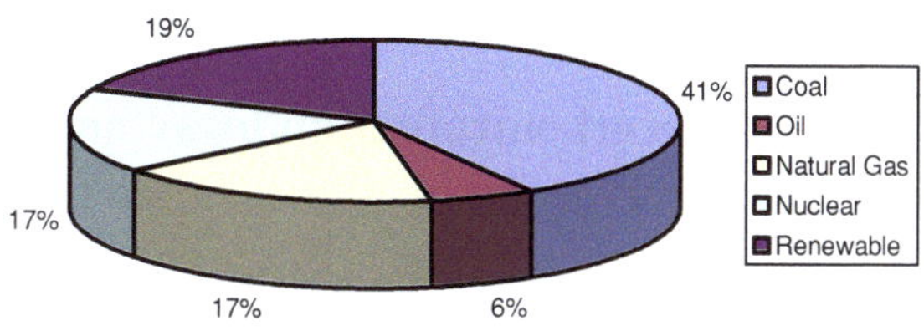

Fig. 1.3 Share of energy production by fuel type (2030 est.)

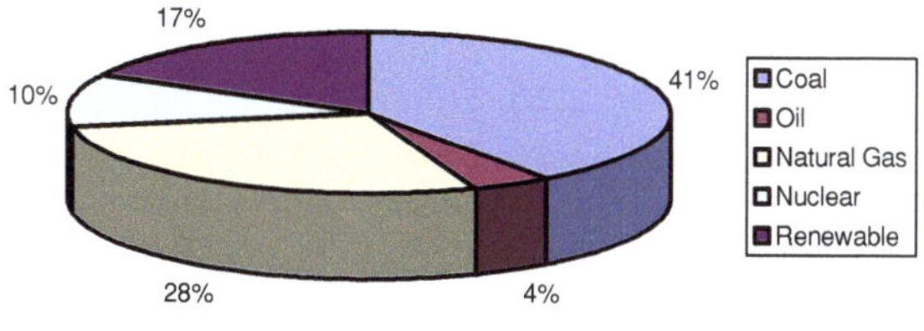

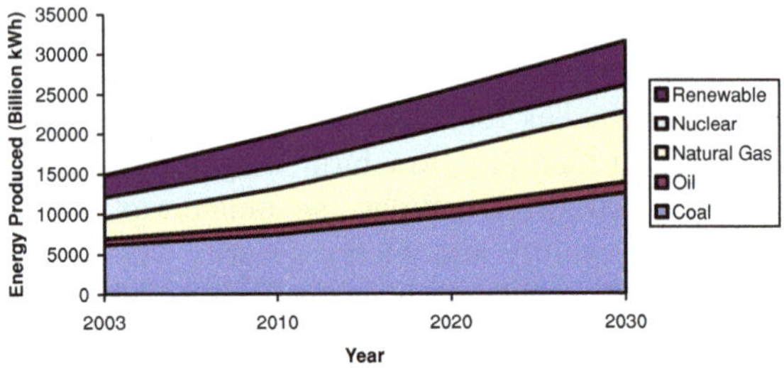

Fig. 1.4 Share of energy fuel type vs. time

Coal will continue to play a significant role. According to EIA, the choice of technology for projected capacity additions is based on the least expensive option available at rates that depend on the current stage of development for each technology. Therefore, EIA estimates assume that the current cost of renewable resources is hindering its potential to play a significant role in the electricity generation market.

Even with conservation and energy efficiency measures, energy demand is projected to double (to 30 TW) by 2050 and more than triple the demand (to 46 TW) by the end of the century. A large driver of this increase in energy demand will be the growing economies of non-industrialized countries and the large populations of the world that are beginning, for the first time, to experience the benefits of living in electrified communities. Currently, 18% of the world's population (OECD[1] countries) is consuming 56% of the world energy and the remaining 82% of the population (non-OECD countries) is consuming 44%. As the lives of people and the economies of the non-OECD countries become more energy intensive, there will be an accelerated growth in energy demand. By 2030, it is projected that the non-OECD countries will be consuming 57% world's energy and OECD countries will be consuming the remainder. This is not due to a reduction in energy use by OECD countries, however. Energy use in OECD countries will be increasing as well and overall energy use is projected to grow by 71% [19].

1.3 The Decarbonization Pathway and the Role of Renewable Energy

An alternative approach to address the anticipated growth in fossil fuel demand and its negative environmental impact would be to continue the use of a least-cost strategy to improve energy efficiency, and minimize pollution without relying on exhaustible primary

[1] OECD Countries Include: United States, Canada, Mexico, Austria, Belgium, Czech Republic, Denmark, Finland, France, Germany, Greece, Hungary, Iceland, Ireland, Italy, Luxembourg, the Netherland, Norway, Poland, Portugal, Slovakia, Spain, Sweden, Switzerland, Turkey, the United Kingdom, Japan, South Korea, Australia, and New Zealand.

energy sources, in the other words, follow the decarbonization-pathway. The recent expansion of natural gas power plants is consistent with this pathway and should continue depending on natural gas pricing and availability. Coal will still play a major role in fueling existing and under-development coal-fired plants. Ultimately, high tech renewable energy technologies, nuclear energy, and clean coal technologies can be utilized to generate low-carbon content fuels and eventually to generate hydrogen to fuel a sustainable global economy.

Over the past 130 years, there has been a transition in the fuels used for heating and electricity production from wood to coal to oil and, lastly, to natural gas. As one moves down this path, it will be noticed that there has been a reduction in the amount of carbon per hydrogen atom in the fuels used. This general trend from use of high carbon fuels to low carbon fuels is called decarbonization. Natural gas has the highest hydrogen to carbon atomic ratio and the lowest CO_2 emissions of all fossil fuels, emitting approximately half as much CO_2 as coal for the same amount of energy. Due to its cost competitiveness and high efficiency, it will also be used to produce an increasing share of the world's electricity. However, as has been demonstrated in the past few years, the heavy reliance on natural gas for electricity generation without a concurrent expansion in its infrastructure has inflated its cost significantly. As a means of offsetting the expected increase of natural gas prices and avoiding the risk of price volatility due to the rapid increase in its usage, it has been proposed, among other measures, to increase the electricity generation from coal-fired plants. The falling prices of coal and the abundance of the coal supply in the certain regions in the world have strengthened this case. However, until the environmental issues that are associated with burning coal and the handling of CO_2 emissions are resolved, coal will remain a non-sustainable source of electricity.

Natural gas offers a bridge to a non-fossil energy future that is consistent with decarbonization. As a fossil fuel, natural gas is a finite resource. Currently, there are no recoverable energy sources with higher hydrogen to carbon atomic ratios than natural gas. In order to avoid moving in the wrong direction down the decarbonization pathway back towards coal and wood as natural gas reserves are depleted, non-fossil energy sources will need to be introduced in the primary energy mix. These non-fossil energy sources include wind, solar and other renewable energy sources. As the natural gas contribution to global energy mix peaks and subsequently declines, carbon-free sources of energy would take over. This would be the end of the decarbonization pathway. A key step towards this goal is the construction of an energy infrastructure that allows hydrogen to be distributed to all end users in much the same way natural gas and electricity are today. In this way, hydrogen can be used to produce heat through combustion and electricity through the use of a combustion engine or fuel cell.

For many years, renewable energy advocates have relentlessly argued that renewable energy has tremendous environmental benefits and can provide a solution to the global warming problem. In addition to and with the ongoing security concerns in the developed countries and the geopolitical problems in oil producing countries, it is becoming clear that there is a need for global as well as national comprehensive energy policies that are based on self-reliance with a significant role for renewable resources to ensure national energy security. However, in spite of their environmental appeal and security role, it is clear that both arguments (environment and energy security) were not enough to spur drastic changes in the forecast for renewable energy's role in future energy policies in many countries worldwide with some exceptions in countries in Europe and the Far East. The cost of renewable energy resources is still high compared with conventional energy systems. The cost of renewable energy technologies has to come down significantly before they can play any significant role in the future of the world's power industry.

1.4 State of Renewable Energy

It is important to have a realistic view of the possibilities as well as the problems associated with the use of renewable energy. The most important challenges for the wide spread adoption of renewable energy technologies can be divided in three categories: economic, technological and social. With regard to economic challenges, renewable energy needs to be generated, stored and utilized in a cost competitive manner. The focus of this study will be to improve the economics of electricity produced from renewable energy resources with hydrogen storage systems.

There are many incentives to improve the technologies associated with and reduce the cost of renewable energy systems. Below are listed some of the most important benefits:

- The operation of renewable energy systems produce no harmful emissions, including NO_x, SO_x, CO and particulates, and therefore have no adverse affects on people's health. Additionally, no green house gases are emitted as a result of their operation.
- Variety and long term viability of methods to produce electricity. As a result of the large number of electricity production technologies, renewable energy is also viewed as a means to enhance energy security. Since electricity can be generated from a variety of renewable sources, any country with a large enough energy resource can increase energy security.
- In many developing areas of the world, the electrical distribution infrastructure needed to provide users with electricity does not exist. The infrastructure requires a large amount of resources and can be expensive. Often, large diesel generators are used to provide electricity in areas with no infrastructure, but this too can be expensive because of fuel requirements and maintenance costs. In these cases, if a significant renewable resource is available, renewable energy systems are a viable alternative.

There are also many problems associated with renewable energy systems. They are:

- Intermittent operation. Because of the resources from which renewable energy is derived, demand loads cannot be met with a high degree of reliability. Energy storage becomes necessary to achieve a high degree of reliability and this storage can be very costly. If renewable energy is ever to compete with conventional, fossil fuel based electricity generation, this problem will need to be addressed.
- Grid stability. Studies by Paynter et al. [21] and Dutton et al. [20] have found that if wind penetration exceeds maximum grid demand by 20–30% then grid stability becomes an issue. Utilization of some form of energy storage will be necessary if higher grid penetration levels are going to be achieved. In this way, energy can be stored when production levels are high and released to the grid later, thereby improving capacity utilization and the economics of the renewable energy system.
- High initial capital costs. As a result of the high initial capital costs of renewable energy systems, smart energy management must be employed to minimize the cost of electricity over the life of the system.

References

1. http://www.earthtrends.wri.org/pdf_library/data_tables/cli4_2003.pdf. Accessed Nov 2005
2. Lewis NS (2005) Basic research needs for solar energy utilization. n.p., Bethesda
3. http://www.worldwildlife.org/climate/basic.cfm. Accessed Nov 2005
4. American Thoracic Society, Committee of the Environmental and Occupational Health Assembly (1996) Health effects of outdoor air pollution. Part 1. Am J Respir Crit Care Med 153:3–50
5. US Environmental Protection Agency (2001) Air quality criteria for particulate matter, vol 2. Publication no. EPA/600/P-99/002bB, Environmental Protection Agency, Research Triangle Park
6. Dockery DW, Pope CA III (1994) Acute respiratory effects of particulate air pollution. Annu Rev Public Health 15:107–132
7. Samet JM, Dominici F, Curriero FC, Coursac I, Zeger SL (2000) Fine particulate air pollution and mortality in 20 US cities. N Engl J Med 343:1742–1749
8. Schwartz J (1994) Air pollution and daily mortality: a review and meta analysis. Environ Res 64:36–52
9. Gauderman WJ, McConnell R, Gilliland F (2000) Association between air pollution and lung function growth in southern California children. Am J Respir Crit Care Med 162:138–1390
10. Bobak M, Leon DA (1999) The effect of air pollution on infant mortality appears specific for respiratory causes in the post-neonatal period. Epidemiology 10:666–670
11. Bobak M (2000) Outdoor air pollution, low birth weight and prematurity. Environ Health Perspect 108:173–176
12. Ha EH, Hong YC, Lee BE, Woo BH, Schwartz J, Christiani DC (2001) Is air pollution a risk factor for low birth weight in Seoul? Epidemiology 12:643–648
13. Ritz B, Yu F (1999) The effect of ambient carbon monoxide on low birth weight among children born in southern California between 1989 and 1993. Environ Health Perspect 107:17–25
14. Ritz B, Yu F, Chapa G, Fruin S (2000) Effect of air pollution on preterm birth among children born in Southern California between 1989 and 1993. Epidemiology 11:502–511

15. Wang X, Ding H, Ryan L, Xu X (1997) Association between air pollution and low birth weight: a community-based study. Environ Health Perspect 105:514–520
16. Woodruff TJ, Grillo J, Schoendorf KC (1997) The relationship between selected causes of postneonatal infant mortality and particulate air pollution in the in the United States. Environ Health Perspect 105:608–612
17. Xu X, Ding H, Wang X (1995) Acute effects of total suspended particles and sulfur dioxides on preterm delivery: a community-based cohort study. Arch Environ Health 50:407–415
18. U.S. Energy Information Association (2006) Annual energy outlook 2006. U.S. Department of Energy, pp 163–174
19. U.S. Energy Information Association (2006) Annual energy outlook 2006. U.S. Department of Energy, pp 83
20. Dutton AG, Bleijs JA, Dienhart H, Falchetta M, Hug W, Prischich D, Ruddell AJ (2000) Experiences in the design, sizing, economics, and implementation of autonomous wind-powered hydrogen production systems. Int J Hydrogen Energy 25:705–722 (Science Direct. Galvin Library, Chicago)
21. Paynter RJH, Lipman NH, Foster JE (1991) The potential of hydrogen and electricity production from wind energy. Energy Research Unit, Rutherford Appleton Laboratory, September 1991

Chapter 2
Renewable Energy Sources and Energy Conversion Devices

Renewable energy sources are those sources that are regenerative or can provide energy, for all practical purposes, indefinitely. These include solar, wind, geothermal, tidal, wave, hydropower and biomass. The status of development, installed capacity, theoretical potential and other considerations are shown in Table 2.1 on the next page.

2.1 Solar Energy

The information contained in this section is based on Duffie and Beckman's Solar Engineering of Thermal Processes, 2nd Edn [8]. For a more comprehensive treatment of the subject of solar engineering and calculations, please refer to the above referenced book.

2.1.1 The Solar Constant

The sun provides energy to the earth in the form of radiation. The radiation emitted by the sun and its spatial relationship to the earth result in a nearly constant intensity of solar radiation at the outer edge of the earth's atmosphere. The amount of energy received by the earth per unit time, based on the average distance between the sun and the earth over the period of a year, is known as the global solar constant, G_{sc}. The generally accepted value of the global solar constant is 1367 W/m^2 (433 Btu/ft^2 h, 4.92 MJ/m^2 h) with an uncertainty of $\sim 1\%$. This value can be used to calculate several values of interest when performing solar calculations.

S. Al-Hallaj and K. Kiszynski, *Hybrid Hydrogen Systems*,
Green Energy and Technology, DOI: 10.1007/978-1-84628-467-0_2,

Table 2.1 Renewable energy resources: status, installed and potential capacity, and environmental, social and aesthetic considerations

	Status of technology	Installed capacity (GW)	Theoretical potential (GW)	Environmental considerations	Aesthetic/social considerations
Solar (Electricity)	Commercial	5.3 (2005) [1]	~82,000		– Land area use
Solar (Thermal)	Commercial	70 (2004) [1]	~82,000		– Land area use
Wind	Commercial	~59 (2005) [2]	2900–7200 [3]	– Low impact on wildlife – Impact on sensitive species possible – Low impact on aquatic species (offshore/near-shore)	– Noisy – Visually intrusive
Geothermal (Electricity)	Commercial	7.7 (2001) [4]	NA	– CO_2, hydrogen sulfide and mercury emissions	NA
Geothermal (Thermal)	Commercial	16.7 (2001) [4]	NA	– Same as geothermal (electrical)	NA
Tidal	Under development	NA	87.4 [5]	– Estuary impact (e.g. sedimentation change) – Impact on migratory fish and birds	NA
Wave	Under development	NA	NA	NA	NA
Hydropower	Commercial	692 (2001) [6]	~2100 [6]	– May disturb aquatic ecosystems – Changes in sedimentation	– Displacement of people – Loss of culture
Biomass (including Wood)	Under development	NA	~41,000	– Incomplete combustion leads to emission of pollutants	– Fuel wood collection burdensome on women and children

2.1.2 Variation of Extraterrestrial Radiation

Variation in the distance between the earth and the sun during the year leads to variations in extraterrestrial radiation of as much as ±3%. The dependence of extraterrestrial radiation on time of year is given by Eq. 2.1.

$$G_{en} = G_{sc}\left(1 + 0.033\,\cos\frac{360n}{365}\right) \tag{2.1}$$

where G_{en} is the extraterrestrial radiation measured on the plane normal to the radiation on the nth day of the year.

The demand for energy is projected to increase to 30 TW by 2050 and 46 TW by 2100 [7]. The earth's solar resource is more than sufficient to meet this demand and the technology to harness this energy is mature enough for deployment, although improvement in the economics is necessary for wide spread adoption.

The wind energy resource can make a significant contribution to energy demand in the near term, is technologically mature and economically attractive. Of the other resources listed, they do not possess sufficient exploitable capacity to meet demand or they are not technologically mature enough for deployment in the short-term. For these reasons, solar and wind energy are the primary focus of this book.

2.1.3 Extraterrestrial Radiation on a Horizontal Surface

To design energy systems that utilize solar radiation, it is often useful to have a means of determining the theoretical solar radiation at the outer edge of earth's atmosphere. The solar radiation incident on a horizontal plane at the outer edge of earth's atmosphere is given by Eq. 2.2.

$$G_0 = G_{sc}\left(1 + 0.033\cos\frac{360n}{365}\right)(\cos\phi\cos\delta\cos\omega + \sin\phi\,\sin\delta) \tag{2.2}$$

where n is the day of the year, ϕ is the latitude, δ is the declination angle, which is the angular position of the sun at solar noon with respect to the plane of the equator, and ω is the hour angle, which is the angular displacement of the sun with respect to the local meridian due to the earth's rotation. Solar noon is time of day when the sun appears the highest in the sky compared to its positions during the rest of the day.

To determine the daily solar radiation on a horizontal surface, H_0, Eq. 2.2 can be integrated from sunrise to sunset to determine the energy per unit area receive by that surface. The result of this integration is Eq. 2.3.

$$H_0 = \frac{24\cdot 3600G_{sc}}{\pi}\left(1 + 0.033\cos\frac{360n}{365}\right)\left(\cos\phi\cos\delta\sin\omega_s\frac{\pi\omega_s}{180}\sin\phi\sin\,\delta\right) \tag{2.3}$$

where ω_s is the sunset hour angle. The sunset hour angle can be determined from Eq. 2.4.

$$\cos \omega_s = -\tan \phi \tan \delta \tag{2.4}$$

Similarly, to determine the hourly solar radiation, I_0, Eq. 2.5 may be used.

$$I_0 = \frac{12 * 3600}{\pi} G_{sc} \left(1 + 0.033 \cos \frac{360n}{365}\right) * \left[\cos \phi \cos \delta (\sin \omega_2 - \sin \omega_1) + \frac{\pi(\omega_2 - \omega_1)}{180} \sin \phi \sin \delta\right] \tag{2.5}$$

2.1.4 Atmospheric Attenuation of Solar Radiation

Solar radiation interacts with the earth's atmosphere in several ways. It may be: (1) reflected back into space, (2) absorbed by gases and particulates in the atmosphere, (3) transmitted directly to the earth's surface or (4) scattered in the atmosphere.

The fraction of radiation reflected back into space is called the albedo and has an annual, latitude-longitude average of 0.35. The reflection is due to reflection from (1) clouds, (2) atmospheric particles and gases and (3) the earth's surface. (4)

The radiation absorbed by the atmosphere causes the warming of the atmosphere. The type and quantity of gases in the atmosphere lead to the attenuation of particular portions of the solar spectrum so that the spectrum at the earth's surface is quite different from that at the outer edge of the atmosphere. X-rays and other very short-wave radiation of the solar spectrum are absorbed by nitrogen and oxygen. Ultraviolet radiation is mainly absorbed by ozone and infrared radiation is mainly absorbed by water vapor and carbon dioxide. After absorption, the resultant wavelength of the spectrum that arrives at the earth's surface is between 0.29 and 2.5 μm (Fig. 2.1).

The radiation can be transmitted directly to the earth's surface or scattered in the atmosphere. Scattering occurs as radiation passes through the atmosphere and interacts with air, water and particulates. The extent of scattering is a function of the degree of particle interactions and the particle size with respect to the wavelength of the radiation. The radiation that is transmitted directly to earth's surface is called direct or beam radiation and the radiation that is scattered is called diffuse radiation. They constitute the radiation that is received by the earth's surface and is used by in solar energy systems.

In summary, the amount of solar radiation available at the earth's surface is a function of time, location (latitude) and also attenuation effects of the earth's atmosphere.

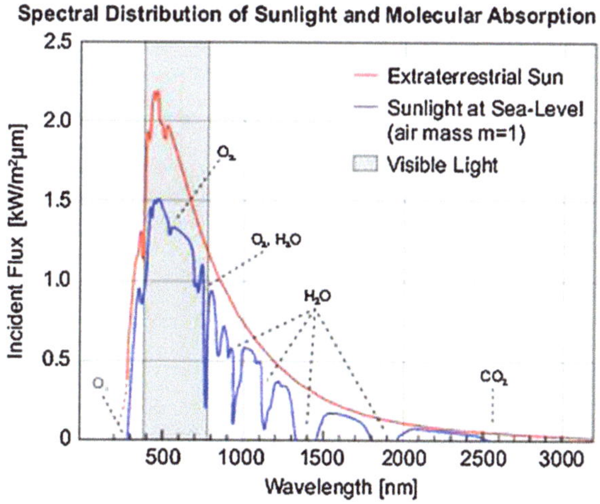

Fig. 2.1 Spectral distribution of extraterrestrial sunlight and sunlight at sea-level [9]

2.1.5 Estimating Monthly Average Solar Radiation

In cases when solar radiation data are not available for a particular location, it is possible to use empirical relationships to estimate radiation values from hours of bright sunshine per day. Equation 2.6 can be used to determine the monthly average daily radiation on a horizontal surface.

$$\frac{\bar{H}}{\bar{H}_0} = a + b\frac{\bar{n}}{\bar{N}} \tag{2.6}$$

where $\bar{H}$ is the monthly average daily radiation on a horizontal surface, $\bar{H}_0$ is the extraterrestrial radiation for the location of interested, a and b are constants that depend on location, $\bar{n}$ is the monthly average daily hours of bright sunshine and $\bar{N}$ is the monthly average number of daylight hours. $\bar{H}_0$ can be calculated from Eq. 2.3. $\bar{N}$ can be calculated from Eq. 2.7 shown below by using the mean day of the month to calculate the declination angle, δ. The mean day for each month is shown in Table 2.2.

$$N = \frac{2}{15}\cos^{-1}(-\tan\phi\tan\delta) \tag{2.7}$$

Values for a and b, the climatic constants, are based on regression analysis of solar radiation data for several geographical locations.

2.1.6 Beam and Diffuse Components of Monthly Radiation

Radiation received at the earth's surface can be split into diffuse and beam radiation. Knowledge of these components is important for calculating the incident

Table 2.2 Average days for months and value for day of year

Month	n for the ith day of month	For the average day of the month	
		Date	n, Day of year
January	i	17	17
February	$21 + i$	16	47
March	$59 + i$	16	75
April	$90 + i$	15	105
May	$120 + i$	15	135
June	$151 + i$	11	162
July	$181 + i$	17	198
August	$212 + i$	16	228
September	$243 + i$	15	258
October	$273 + i$	15	288
November	$304 + i$	14	318
December	$334 + i$	10	344

radiation on surfaces that have an orientation that differs from those surfaces for which data is available. It is also important to know the beam component of the total radiation to determine the long-term performance of concentrating solar collectors.

Erbs et al. developed monthly average correlations for the diffuse portion of solar radiation [10]. The correlation is dependent on the monthly average clearness index, $\bar{K}_T$, which is defined as $\bar{H}/\bar{H}_0$. The equations for the correlation are shown below. $\bar{H}_d$ is the monthly average diffuse component of radiation:

For $\omega_S \leq 81.4°$ and $0.3 \leq \bar{K}_T \leq 0.8$

$$\frac{\bar{H}_d}{\bar{H}} = 1.391 - 3.560\bar{K}_T + 4.189\bar{K}_T^2 - 2.137\bar{K}_T^3 \tag{2.8}$$

and for $\omega_S > 81.4°$ and $0.3 \leq \bar{K}_T \leq 0.8$

$$\frac{\bar{H}_d}{\bar{H}} = 1.311 - 3.022\bar{K}_T + 43.427\bar{K}_T^2 - 1.821\bar{K}_T^3 \tag{2.9}$$

2.1.7 Radiation on Sloped Surfaces

To determine the radiation on sloped surfaces when only the total radiation is known, the directions from which the diffuse and beam components reach the surface being studied must be determined.

The solar radiation models for determining incident radiation on a sloped surface are based on measured global irradiance on a horizontal surface. On a clear day, the diffuse radiation is composed of three parts: the isotropic component, which is received uniformly from the entire sky dome; the circumsolar diffuse component, which results from scattering of solar radiation and is concentrated in the area of the sky around the sun; and the horizon brightening component, which is concentrated near the horizon and is apparent in clear skies. The differences among the solar radiation models are in the way they treat the three parts of the diffuse radiation.

The total incident radiation on a sloped surface can be determined using Eq. 2.10.

$$I_T = I_b R_b + I_{d,iso} F_{c-s} + I_{d,cs} R_b + I_{d,hz} F_{c-hz} + I \rho_g F_{c-g} \tag{2.10}$$

I_b is the incident beam radiation on the tilted surface. R_b is ratio of beam radiation on a tilted plane to that on the plane of measurement and can be calculated using Eq. 2.11. $I_{d,iso}$ is the incident diffuse radiation due to isotropic part of the diffuse radiation, F_{c-s} is the view factor of the tilted surface to the sky, $I_{d,hz}$ is diffuse radiation due to horizon brightening, F_{c-hz} is the view factor of the tilted surface to the horizon, I is the total radiation incident on a horizontal surface, ρ_g is the ground reflectance and F_{c-g} is the view factor of the tilted surface to the ground.

$$R_b = \frac{\cos\theta}{\cos\theta_z} \tag{2.11}$$

θ is called the angle of incidence and is the angle between the beam radiation on a surface and the normal to that surface. θ_z is called the zenith angle and is the angle of incidence of beam radiation on a horizontal surface.

Equations 2.12 and 2.13 are helpful relationships for determining the values of θ and θ_z, respectively.

$$\begin{aligned}\cos\theta = {} & \sin\delta \sin\phi \cos\beta - \sin\delta \cos\phi \sin\beta \cos\gamma \\ & + \cos\delta \cos\phi \cos\beta \cos\omega + \cos\delta \sin\phi \sin\beta \cos\gamma \cos\omega \\ & + \cos\delta \sin\beta \sin\gamma \sin\omega \end{aligned} \tag{2.12}$$

β is called the sloped and is the angle between the surface of interest and the horizontal plane. γ is called the surface azimuth angle and is the angular deviation from the local meridian of the projection on a horizontal plane of the normal to the surface. ω is called the hour angle and is the angle of the sun east or west of the local meridian.

$$\cos\theta = \cos\theta_z \cos\beta + \sin\theta_z \sin\beta \cos(\gamma_s - \gamma) \tag{2.13}$$

γ_s is called the solar azimuth angle and is the angle between the projection of the beam radiation on the horizontal plane and south. It can be determined using Eq. 2.14 through Eq. 2.16.

$$\gamma_s = C_1 C_2 \gamma_s' + C_3 \left(\frac{1 - C_1 C_2}{2}\right) 180 \tag{2.14}$$

$$\sin \gamma_s' = \frac{\sin \omega \cos \delta}{\sin \theta_z} \tag{2.15}$$

$$\begin{aligned} C_1 &= 1 \quad \text{if } |\omega| \leq \omega_{ew} \quad \text{or } -1 \text{ if } |\omega| \geq \omega_{ew} \\ C_2 &= 1 \quad \text{if } |\phi - \delta| \geq 0 \quad \text{or } -1 \text{ if } |\phi - \delta| \leq 0 \\ C_3 &= 1 \quad \text{if } \omega \geq 0 \quad \text{or } -1 \text{ if } \omega \leq 0 \end{aligned}$$

$$\cos \omega_{ew} = \frac{\tan \delta}{\tan \phi} \tag{2.16}$$

ω_{ew} is the hour angle when the sun is due east or west.

The solar radiation on tilted surfaces can be determined with using several different solar radiation models using measured data for a horizontal surface. The three models described in this section are:

1. Isotropic diffuse model I
2. Hay and Davies model
3. HDKR model

2.1.8 The Isotropic Diffuse Model I

The Isotropic Diffuse Model, derived by Liu and Jordan, includes three components to determine the total irradiance: beam, isotropic diffuse and ground reflectance. The third and fourth terms from Eq. 2.10 are considered to be zero since all diffuse radiation is assumed to be isotropic. F_{c-s} is given by (1 + cosβ)/2. F_{c-g} is given by (1 − cosβ)/2. Equation 2.10 becomes Eq. 2.17.

$$I_T = I_b R_b + I_d \left(\frac{1 + \cos \beta}{2} \right) + I \rho_g \left(\frac{1 - \cos \beta}{2} \right) \tag{2.17}$$

2.1.9 The Hay and Davies Model

The Hay and Davies model differs from the isotropic model in that it provides an estimate of the fraction of diffuse radiation that is circumsolar and assumes that it is from the same direction as the beam radiation. Therefore, the third term in Eq. 2.10 is not zero. The diffuse radiation on the tilted surface is given by Eq. 2.18.

$$I_{d,T} = I_d \left[(1 - A_i) \left(\frac{1 + \cos \beta}{2} \right) + A_i R_b \right] \tag{2.18}$$

A_i is called the anisotropy index and is defined by Eq. 2.19.

$$A_i = \frac{I_b}{I_0} \tag{2.19}$$

Equation 2.10 then becomes Eq. 2.20.

$$I_T = (I_b + I_d A_i) R_b + I_d (1 - A_i) \left(\frac{1 + \cos\beta}{2} \right) + I\rho_g \left(\frac{1 - \cos\beta}{2} \right) \tag{2.20}$$

2.1.10 The HDKR Model

The HDKR model is modified form of the Hay and Davies Model. It contains a horizon brightening term. The diffuse radiation on the tilted surface is given by Eq. 2.21.

$$I_{d,T} = I_d \left[(1 - A_i) \left(\frac{1 + \cos\beta}{2} \right) \left[1 + f \sin^3 \left(\frac{\beta}{2} \right) \right] + A_i R_b \right] \tag{2.21}$$

F is defined by Eq. 2.22.

$$f = \sqrt{\frac{I_b}{I}} \tag{2.22}$$

With these modified terms included, Eq. 2.21 becomes Eq. 2.23.

$$I_T = (I_b + I_d A_i) R_b + I_d (1 - A_i) \left(\frac{1 + \cos\beta}{2} \right) \left[1 + f \sin^3 \left(\frac{\beta}{2} \right) \right] + I\rho_g \left(\frac{1 - \cos\beta}{2} \right) \tag{2.23}$$

2.2 Photovoltaic Cells

Photovoltaic (PV) cells are devices that absorb light and convert this light directly into electricity. A picture of a solar cell is shown in Fig. 2.2. The solar cell has a dark area, which is usually silicon, and thin silver areas. The silicon absorbs the sunlight and a voltage is generated between the front and back of the cell. The thin sliver areas are called front contact fingers and are used to create an electrical contact to the front of the cell. The back of the cell is a solid metal layer that reflects light back up through the cell and provides an electrical contact on the back side. Commercially available cells have a 25–30 year lifetime.

Solar cells are strung together in various serial or parallel configurations to achieve desired electrical characteristics and assembled into modules.

Fig. 2.2 A PV single cell [11]

Fig. 2.3 Mono-crystalline PV module [12]

Modules consist of solar cells sealed between two pieces of glass and framed. A picture of PV module is shown in Fig. 2.3. The module protects cells from mechanical damage from handling and the environmental and provides the end user with a robust package that can easily be connected to other modules. A system of connected modules is called an array. The array, in conjunction with electrical conditioning equipment constitutes a complete photovoltaic system. It can be used to meet a user's electrical demand load.

2.2.1 PV Cell Electrical Performance

Solar cells are characterized by their open-circuit voltage, short-circuit current and I–V (current–voltage) curve. The open-circuit voltage of the cell is the voltage

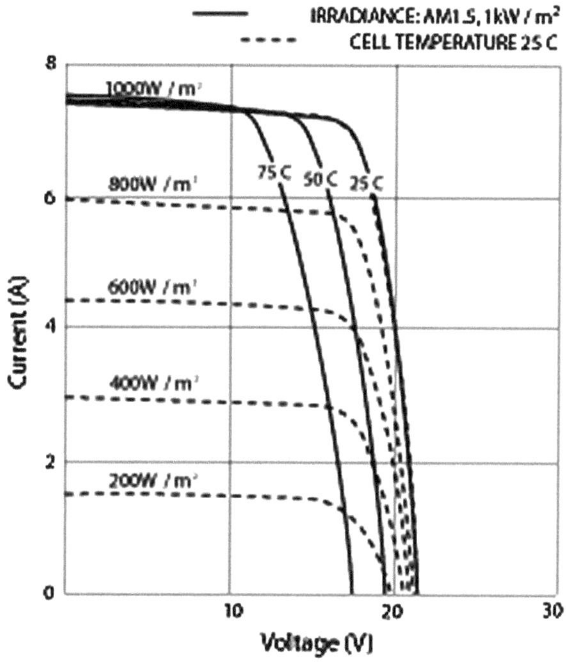

Fig. 2.4 I–V curve of PV module under different irradiances and at different temperatures [13]

across the cell when a very high resistance load is connected to the cell. In this case no current is being drawn from the cell. If the load resistance is reduced to zero, the cell is short-circuited and the resultant current in this situation is the short-circuit current. The short-circuit current is directly proportional to the light falling on the cell. The I–V curve is a plot of the current vs. the voltage of the cell as the load is varied. An example of an I–V curve for various ambient temperatures and insolation levels is shown in Fig. 2.4. The power being generated by the solar cell is the product of the voltage and current at any point on the curve. The point on the curve at which the product is greatest is called the maximum power point. The resistance of the load at this point is equal to the cell's internal resistance.

Cell performance is determined under a specified set of standard conditions. Under these conditions, the ambient temperature is 25°C and the power density of the incident solar radiation is 1 kW/m^2 with a spectrum that is equivalent to sunlight that has passed through the atmosphere when the sun is at a 42° elevation from the horizon.

2.2.2 *Types of PV Cell*

There are three main categories of PV cells: (1) inorganic, (2) organic and (3) photoelectrochemical. Of the inorganic type, three sub-types are commercially available: mono-crystalline, multi-crystalline and amorphous. Mono-crystalline cells are constructed from single crystal silicon ingots by slicing the ingots as is done in microchip fabrication, while multi-crystalline cells are constructed by evaporating coatings onto a substrate. Because grain boundaries exist between crystals in the multi-crystalline cells, electrons can cross at the boundaries which increase losses and reduce efficiency with respect to mono-crystalline cells. Cells using amorphous silicon (a-Si) are of interest because they can be manufactured on continuous lines and require less silicon than the other PV cell types described. The main problem with a-Si cells currently is that they are less efficient than multi-crystalline and mono-crystalline silicon cells. But, because of their lower manufacturing costs, they are able to achieve a lower cost/watt-peak. Typical efficiencies of commercially available PV panels range from 10 to 17%.

2.2.3 Inorganic Solar Cell Operation

Photons from the sun can be captured and used directly to produce electricity through the use of a PV cell, which can be made of several semiconductor materials. An explanation of bandgap energy is useful for understanding photon absorption. The bandgap energy is the difference in energy between the top valence band (lower energy) and bottom conduction band (higher energy), as shown in Fig. 2.5. The bands refer to the electron energy ranges for electrons in different electron orbitals. The electrons in the conduction band can be used to create an electrical current. When a photon hits a piece of silicon or another type of semiconductor, several events may occur. If the photon energy is lower than the bandgap energy of the silicon semiconductor, it will pass through the silicon. If the photon energy is greater than the bandgap energy, it will be absorbed.

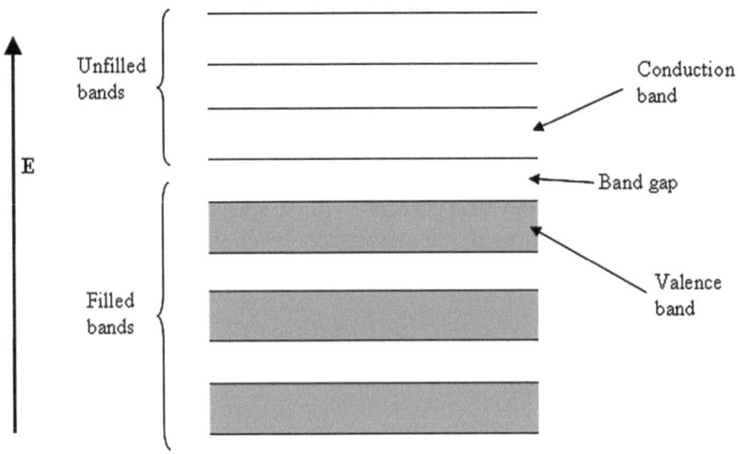

Fig. 2.5 Schematic of electron bands [14]

When a photon is absorbed, its energy can be given to an electron in the valence band, an electron–hole pair created and heat may be generated. The energy given to the electron excites it and moves it into the conduction band. Now it is free to move around within the semiconductor. The band that the electron was previously a part of now has one less electron. This is known as a hole. The presence of a missing covalent bond allows the bonded electrons of surrounding atoms to move into the hole leaving another hole behind. By this mechanism, a hole can move through the semiconductor opposite the direction that the electrons flow. As a result, moving electrons and holes are created and move to opposite sides of the cell structure. This freedom of movement that occurs as a result of the absorption of the photon is what allows a charge differential (voltage) to be created within the cell. By connecting an external circuit to the two sides of the silicon cell, electrons can recombine with holes and this flow of electrons via this external circuit can be utilized to do work.

To construct an actual cell, an n-type semiconductor (one with excess electrons that can donate electrons) and a p-type semiconductor (one deficient in electrons that can accept electrons) must be placed in physical contact with one another. The semiconductors can be modified by adding impurities, known as dopants, to the Si material. Depending on the number of electrons possessed by the dopant, an n-type or p-type semiconductor can be formed.

2.2.4 Organic Solar Cell Operation

Organic solar cells also operate with junctions, but the n-type and p-type semiconductors are organic compounds, and the junction between the n- and p-type materials does not produce an electric field. It has a different function than the inorganic p–n junctions. When electrons and holes are produced upon absorption of light, the electrons and holes become bound to one another to form electron–hole pairs called excitons. The excitons have no net electrical charge and cannot carry current. They must be broken apart in order to produce the free electrons and holes required to generate a current. This is the function of the junction between the n- and p-type organic compounds. When the excitons diffuse to this region of the cell, they split apart and produce the required free electrons and holes.

2.2.5 The Current State and Future of Solar Cells

Solar cell technologies are divided into Generation I, II and III cells. Generation I cells include single-crystal and multi-crystal Si solar cells. Generation II cells include those that involve the use of several types of thin films including both inorganic and organic materials. Generation III cells are denoted by their ability to operate above the 32% thermodynamic efficiency limit known as the

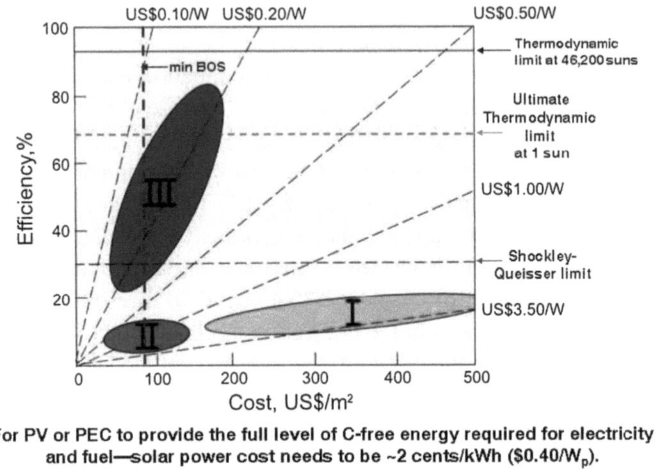

Fig. 2.6 Cost and efficiency of generation I, II and III PV cells [15]

Shockley-Queisser limit. How is this possible? There are two suggested ways. One assumption made in the original calculations of the above limit is that the energy of an absorbed photon above the bandgap energy becomes heat. If this assumption is not made, the thermodynamic increases to 65%. This can only be accomplished if the material that absorbs the photon is able to utilize the excess photon energy. The other way to accomplish this is through the use of multi-junction cells. Multi-junction cells utilize several materials together with different bandgaps that are able to utilize photons with different energy levels. These cells have been shown to operate at efficiencies higher than those of single-crystal Si cells.

Figure 2.6 shows efficiency plotted against cost (US$/m^2) for Generation I, II and III cells. Dashed lines show the cost per Wp and the Shockley-Queisser limit and thermodynamic limits at 1 sun and 46,200 suns are also shown.

Figure 2.7 shows the efficiency of different PV technologies plotted against time. The different colored lines and tick marks correspond to different PV technologies. The names at each tick mark are the labs or companies that achieved the given efficiency.

2.2.6 Solar Concentrators and Trackers

In order to reduce the cost of PV systems, solar concentrators may be utilized. A solar concentrator focuses light on a solar cell at intensities several times higher than that provided by a single sun. Several types of concentrators exist including parabolic, Fresnel and Winston. The benefit of such an approach is that, since the solar cell current is proportional to the intensity of the light incident to the cell, the solar cell should provide a higher current. By substituting inexpensive optics for expensive semiconductor material, system costs are reduced.

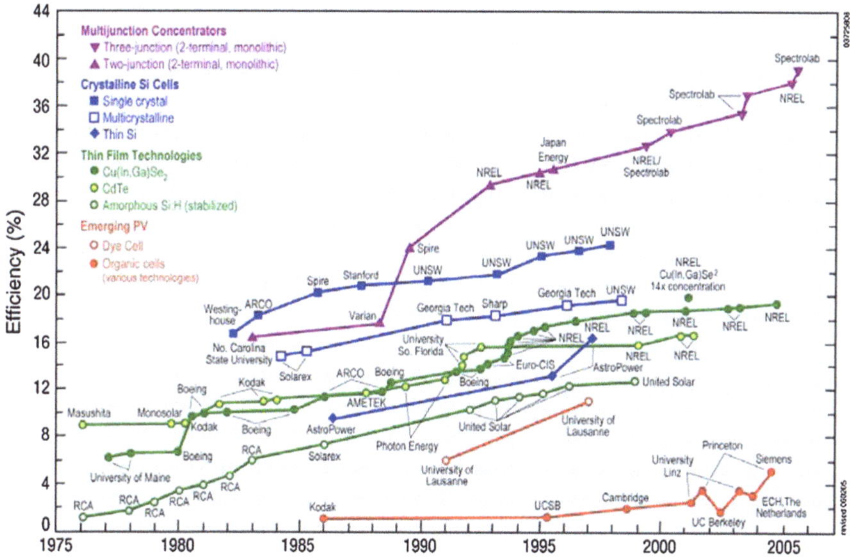

Fig. 2.7 PV efficiencies from 1976 to 2004 [16]

Such a system is not without its problems. First, because of the increased light intensity, the cell begins to overheat and some passive or active means of heat management is required. This will increase costs. Depending on the geographical location of the system, the use of concentrators may not make sense. Additionally, solar cell performance will decrease with an increase in cell temperature.

Second, the typical material used to seal the PV cells between glass, EVA, will degrade at high temperatures. More expensive encapsulation materials, such as silicon, will be required and this too will increase costs.

Lastly, for concentrators to be most effective, a tracker system will be required. Two types of tracking systems exist: single axis and double axis. Single axis systems track the sun as it moves throughout the day (azimuthal tracking). Double axis systems follow both the daily movement of the sun across the sky and the annual change in the sun's position. For a flat-plate collector, which a PV module is, the maximum average gain from using a two axis system (less than a factor of 2) would rarely justify the extra cost of the tracking system, which currently costs more than simply doubling the collector area [17].

2.2.7 Economics

Photovoltaic cells incur no fuel expenses, but they do involve a capital cost. The cost for the electricity produced by the cell is determined based on the capital cost of the PV module and the total electrical energy generated over the module's lifetime. In addition to module costs, a PV system also has costs associated with the power conditioning and energy storage components of the system. These are called balance of system costs, and they are currently in the range of $250/m^2 for Generation I cells. Therefore, the total cost of present PV systems is about $6/Wp. A quick rule of thumb to convert the $/Wp cost figure to $/kWh follows the relationship: $1/Wp $\sim$ $0.05/kWh. This calculation leads to a present cost for grid-connected (no storage) PV electricity of about $0.30/kWh.

2.3 Wind Energy

2.3.1 Wind Generation

Wind is caused by air flowing from high pressure to low pressure regions of the atmosphere. There are two causes for this variation in pressure: (1) the heating of the atmosphere by the sun and (2) the rotation of the earth.

The warming effect of the sun varies with latitude and with the time of day. Warmer air is less dense than cooler air and rises above it so the pressure above the equator is lower than the pressure above the poles. This phenomenon results in a convective current that moves the air higher in the atmosphere at the equator, then toward the poles where the air cools and falls back towards the earth's surface, and finally returns to the equator.

As the earth spins on its axis, it drags the atmosphere with it. The air higher up in the atmosphere is less affected by this drag effect. Instead of traveling in a straight line, the path of the moving air veers to the right. The result of the phenomena described is that the wind circles in a clockwise direction towards the area of low pressure in the Northern hemisphere and counter-clockwise in the Southern hemisphere.

2.3.2 Wind Data Collection and Siting

Sites that may be attractive to wind development are selected using a wind map and validated with wind measurements. A wind map is constructed using meteorological wind data, which is usually not sufficient to accurately site a large wind power project. After preliminary identification, data collection equipment (anemometer) is used to record the wind speed and direction for several locations at a given site, usually for a year. A higher resolution wind map of the area of interest is then constructed to identify the best location for turbines. This process is known as micro-siting. A good location would have a near constant flow of non-turbulent wind throughout the year without too many sudden bursts of wind.

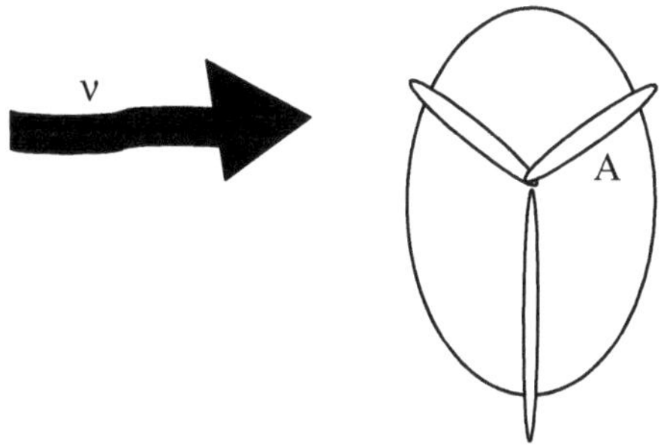

Fig. 2.8 Diagram showing air (fluid) with speed, v, and density, ρ, moving perpendicular to wind turbine blades with swept area, A

Wind farms often have many turbines installed. Since each turbine extracts some of the energy of the wind, it is important to provide adequate spacing between turbines to prevent interference and reduced energy conversion. As a general rule, turbines are spaced three to five rotor diameters apart perpendicular to the prevailing wind and five to ten rotor diameters apart in the direction of the prevailing wind

The power of the wind passing perpendicularly through a given area, shown in Fig. 2.8, is determined by Eq. 2.24 shown below.

$$P = \frac{1}{2}\rho A v^3 \tag{2.24}$$

where

$P =$ power
$\rho =$ density of the fluid
$A =$ the area through which the fluid is flowing
$v =$ the wind speed

The theoretical maximum amount of energy that can be extracted by a wind turbine was determined to be 59.3% by Albert Betz [18].

2.3.3 Wind Turbine Types and Operation

Several types of wind turbine designs exist, though the dominant design is a horizontally-mounted propeller type shown in Fig. 2.9.

These types of turbines typically have two or three blades that are evenly distributed around the axis of rotation. These blades are designed to rotate at a specific angular velocity and torque. The mechanical energy produced can then be further converted to other forms of energy depending on the application (Fig. 2.9).

To covert the mechanical energy to electrical energy, a generator is coupled to the shaft of the propeller. The generator can be designed to operate at a fixed or variable angular velocity. For those generators designed to work at a fixed angular velocity, a transmission will be used to change the angular velocity from that of

Fig. 2.9 Photo of horizontal axis wind turbine [19]

the propeller to that required by the generator. This type of generator will output the electrical energy as an AC current with constant frequency (usually that of the electrical grid to which it is connected). Alternatively, the energy converter may operate in a variable frequency mode. The electrical energy of this type of system would be output as an AC current with varying frequency [20].

2.3.4 Determining Wind Turbine Power Output

To determine the power generated in a particular hour by a given turbine, three calculations are required. First, the wind speed data must be corrected to compensate for the difference between the height of the wind anemometer that was used to collect the wind speed data and the hub height of the turbine being considered in the calculations. The hub height is the height of the turbine as measured from the ground where the turbine is installed to the center of rotation of the turbine blades. The wind speed profile is shown in Fig. 2.10.

A logarithmic or power law relationship can be used to determine the wind speed at the turbine's hub height as shown in Eq. 2.25 and 2.26 respectively.

$$\frac{v(z_{hub})}{v(z_{anem})} = \frac{\ln(z_{hub}/z_0)}{\ln(z_{anem}/z_0)} \tag{2.25}$$

Fig. 2.10 Logarithmic wind speed profile with surface roughness length = 0.15

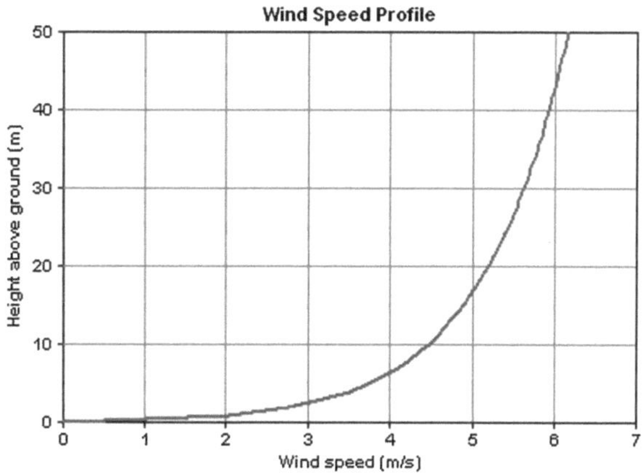

where

$v(z_{hub})$ = wind speed at hub height
$v(z_{anem})$ = wind speed at anemometer height
z_{hub} = hub height of the wind turbine
z_{anem} = height of wind anemometer
z_0 = surface roughness length

$$\frac{v(z_{hub})}{v(z_{anem})} = \left(\frac{z_{hub}}{z_{anem}}\right)^{\alpha} \tag{2.26}$$

Where: α = the power law exponent (Note: Research in fluid mechanics has shown that its value is equal to 1/7 for turbulent flow over a flat plate. Wind speed researchers, however, have found that the exponent depends on temperature, season, terrain roughness, and several other factors.)

Second, a power curve as shown in Fig. 2.11 for the turbine is used to determine the power output of the turbine given the corrected wind speed. The power curve is a plot of the power output for a turbine plotted against wind speed. The turbine shown in this example is a WES18 80 kW turbine [21].

Lastly, the power captured from the wind depends not only on the wind turbine and the wind speed, but also on the air density. Thus, the power output determined above in the second step is multiplied by the air density ratio, which is the air density at the location of interest divided by the air density at standard conditions. This ratio corrects for the altitude of the location of interest and is calculated as shown below in Eq. 2.27.

$$\frac{\rho}{\rho_0} = \left(1 - \frac{Bz}{T_0}\right)^{\frac{g}{RB}} \left(\frac{T_0}{T_0 - Bz}\right) \tag{2.27}$$

where

ρ = air density (kg/m^3)

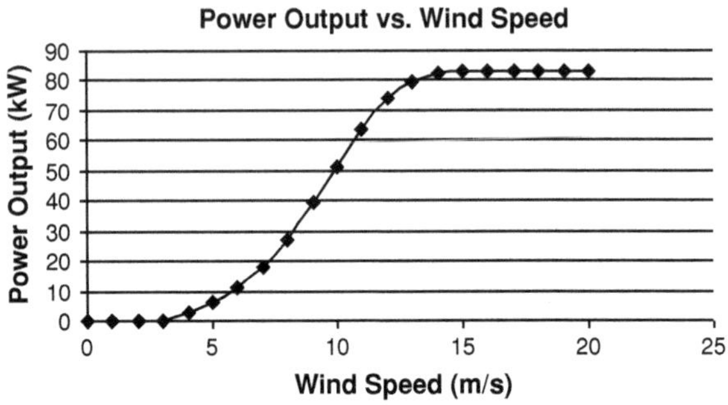

Fig. 2.11 Power curve for WES 18 wind turbine [21]

ρ_0 = standard air density (kg/m^3)
B = lapse rate (0.00650 K/m)
z = altitude (m)
T_0 = standard temperature (K)
g = acceleration due to gravity (9.81 m/s^2)
R = gas constant (287 J/kg · K)

References

1. Trends in Photovoltaic Applications. International Energy Agency (2006)
2. Wind Energy Association- Statistics." World Wind Energy Association. 6 Mar 2006. World Wind Energy Association. 9 Jan 2007. http://www.wwindea.org/home/index.php?option=com_content&task=blogcategory&id=21&Itemid=43
3. Archer CL, Jacobson MZ (2004) Evaluation of global wind power. J Geophys Res. 9 Jan 2007. http://www.stanford.edu/group/efmh/winds/global_winds.html
4. Carella R (2001) World energy council survey of energy resources. World Energy Council. 9 Jan 2007. http://www.worldenergy.org/wec-geis/publications/reports/ser/geo/geo.asp
5. Craig J (2001) World energy council survey of energy resources. World Energy Council. 9 Jan 2007. http://www.worldenergy.org/wec-geis/publications/reports/ser/tide/tide.asp
6. Lafitte R (2001) World energy council survey of energy resources. World Energy Council. 9 Jan 2007. [Craig J (2001) World energy council survey of energy resources. World Energy Council 9 Jan 2007]
7. Nault RM (2005) comp. Basic Research Needs for Solar Energy Utilization. Basic Energy Sciences, U.S. DOE. Argonne National Laboratory, Argonne, IL, 2005. 3. http://www.sc.doe.gov/bes/reports/files/SEU_rpt.pdf
8. Duffie JA, Beckman WA (1991) Solar energy engineering, 2nd edn. Wiley, London, pp 1–146
9. Digital image [Spectral Distribution of Sunlight]. London Metropolitan University. http://www.learn.londonmet.ac.uk/packages/clear/visual/daylight/sun_sky/images/solar_radiation.png
10. Erbs DG, Klein SA, Duffie JA (1982) Estimation of the diffuse radiation fraction for hourly, daily and monthly-average global radiation. Sol Energy 28:292–302
11. Digital image [Picture of Siemen Cell]. Energy Center of Wisconsin. http://www.wisconsun.org/images/siemen_cell.jpg
12. Digital image [Sharp PV Module]. http://www.hanseatic-trade-company.de/images/sharp-modul.jpg

13. Digital image [PV Module I-V Curver]. http://www.kyocerasolar.com/images/SI_irradiance.gif
14. Digital image [Electronic Band Diagram]. http://en.wikipedia.org/wiki/Image:Electronic_band_diagram.svg
15. Nault RM (2005) comp. Basic Research Needs for Solar Energy Utilization. Basic Energy Sciences, U.S. DOE. Argonne National Laboratory, Argonne, IL 2005. 14. http://www.sc.doe.gov/bes/reports/files/SEU_rpt.pdf
16. Nault RM (2005) comp. Basic Research Needs for Solar Energy Utilization. Basic Energy Sciences, U.S. DOE. Argonne National Laboratory, Argonne, IL, 2005. 18. http://www.sc.doe.gov/bes/reports/files/SEU_rpt.pdf
17. Sorensen B (2000) Renewable energy, 2nd edn. Academic Press, London, pp 382–385
18. Sorensen B (2000) Renewable energy, 2nd edn. Academic Press, London, p 345
19. Digital image [Large Windmill]. http://www.leicageosystems.com/news/images/windmill.jpg
20. Sorensen B (2000) Renewable energy, 2nd edn. Academic Press, London, p 448
21. "Wind Turbine Manufacturer- WES BV." WES Wind Turbines

Chapter 3
Hydrogen Production, Storage and Fuel Cells

3.1 Hydrogen Production Methods

Hydrogen can be produced through thermal, electrolytic, or photolytic processes using fossil fuels, biomass, or water as a feedstock. Photolytic processes will not be covered here.

Thermal processes used to produce hydrogen from methane include steam methane reforming (SMR), partial oxidation (POX) and autothermal reforming (ATR), which combines the SMR and POX processes. When heavy oils or coal is used, the gasification process is commonly used.

3.1.1 Steam Methane Reformation

Steam reforming is a widely used method of producing syngas, a mixture of hydrogen and carbon monoxide, from methane. The typical feedstock is natural gas, which comes in several varieties including dry, wet, sweet and sour gas. These designations refer to the composition of the gas. Dry gas is mostly methane, whereas wet gas contains higher hydrocarbons. Sweet gas has little hydrogen sulfide, whereas sour gas contains higher levels of hydrogen sulfide.

Steam reforming of methane typically consists of four steps: (1) Hydrogen sulfide and other sulfur compounds are removed to prevent catalyst poisoning; (2) Pre-reforming is used to protect against carbon formation during the main reforming step and also reduce the amount of steam required; (3) Primary reforming in which steam and heat are supplied to allow the reaction to proceed over a nickel catalyst at 700–830°C; (4) A secondary reformer utilizes air to produce heat through combustion reactions to bring the temperature to ~1300°C and convert most of the remaining methane to syngas.

The desulfurization process is an exothermic process that is typically carried out in a packed bed reactor. The sulfur compounds are adsorbed by the packed bed, which is usually ZnO.

S. Al-Hallaj and K. Kiszynski, *Hybrid Hydrogen Systems*,
Green Energy and Technology, DOI: 10.1007/978-1-84628-467-0_3,

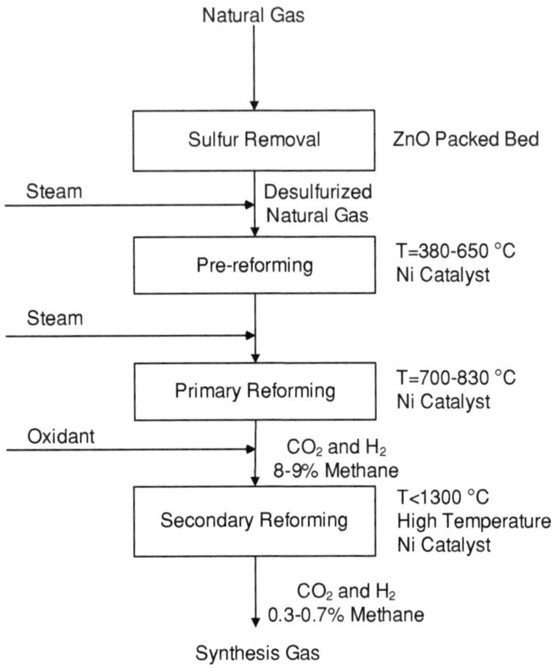

Fig. 3.1 Steam reforming of natural gas process block diagram [3]

The pre-reforming step is carried out at a temperature range of 380–650°C over a catalyst. The process is endothermic for natural gas and exothermic for heavier feedstocks. This process further removes any sulfur left from the previous step.

The primary steam methane reforming step is an endothermic process $\left(\Delta H^{\circ}_{298} = +206\,\mathrm{kJ\,mol^{-1}}\right)$, which involves reacting methane with steam (Eq. 3.1).

$$CH_4 + H_2O \leftrightarrow CO + 3H_2 \tag{3.1}$$

In the secondary reformer, two steps take place. First, an oxidant, either air or oxygen is supplied to the reaction environment. The oxidant reacts with the incoming gas from the primary reformer in an exothermic reaction to form water and raise the temperature. In the second step, a nickel catalyst bed is used to catalyze the endothermic reforming reaction. This process reduces the methane content of the product gas of the primary reformer step from 8–9% to 0.3–0.7% (Fig. 3.1).

3.1.2 Water–Gas Shift Reaction

After the syngas is produced, it is cooled and more hydrogen is produced through the water gas shift (WGS) reaction (Eq. 3.2). The WGS reaction involves reacting carbon monoxide with steam over a catalyst to produce hydrogen and carbon dioxide $\left(\Delta H^{\circ}_{298} = -41\,\mathrm{kJ\,mol^{-1}}\right)$.

$$CO + H_2O \rightarrow CO_2 + H_2 \quad (3.2)$$

3.1.3 Partial Oxidation of Methane

Partial oxidation (Eq. 3.3) of methane is an exothermic process $\left(\Delta H^{\circ}_{298} = -36\,\mathrm{kJ\,mol^{-1}}\right)$ whereby a fuel is reacted with oxygen to produce a mixture of carbon monoxide and hydrogen.

$$CH_4 + 1/2O_2 \rightarrow CO + 2H_2 \quad (3.3)$$

Problems associated with this process include catalyst degradation due to carbon deposition and explosion risks in during mixing of the reactant gases [1–4].

3.1.4 Autothermal Reformation of Methane

Autothermal reformation, also known as oxy-steam reforming, combines SMR and POX. There are several benefits to this approach. The steam can be used to convert some of the carbon monoxide to carbon dioxide via the WGS reaction. It can also reduce risk of explosion and reduce carbon deposition on the catalyst. When the proper mixture of fuel, air and steam are supplied, the partial oxidation reaction supplies all the heat required to drive the steam reforming reaction.

3.1.5 Hydrogen Production from Heavy Oil and Coal

The processes used to produce hydrogen from heavy oil and coal feedstock are similar. In the case of oil, the process used is called partial oxidation. This process converts petroleum feedstock into syngas, CO_2, CH_4 and H_2S if sulfur is present in the feedstock. The oil is exposed to steam and controlled amounts of oxygen at 1200–1500°C and pressure of 30–80 bars. There are three steps in this process: (1) Steam is used to reduce the size of the hydrocarbon chains (known as cracking); (2) Substoichiometric amounts of oxygen oxidize the oil into syngas; (3) Carbon particulates react with CO_2 and steam to form syngas. If sulfur is present, it is removed before the WGS reaction as sulfur can poison the catalyst used for the WGS reaction (Fig. 3.2).

Coal can also be used as a feedstock to produce syngas. The processes involved include pretreatment, primary gasification, secondary gasification and

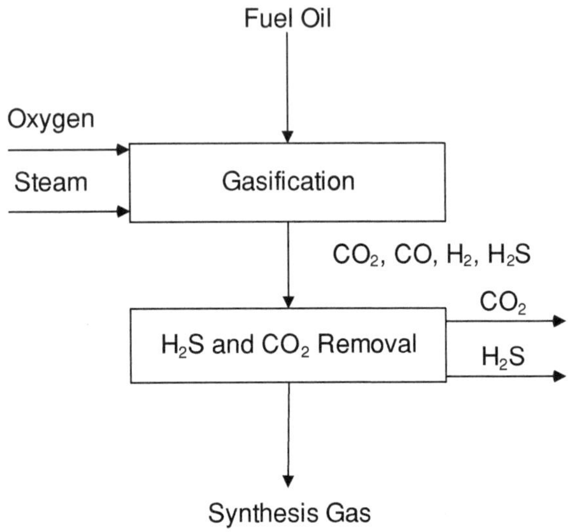

Fig. 3.2 Partial oxidation of fuel oil block diagram [3]

shift conversion. During pretreatment, oxygen is introduced to remove compounds that would otherwise cause the coal to agglomerate in the gasifier. The primary gasifier produces synthesis gas, CO_2, H_2O, CH_4, N_2, char (primarily carbon) and other compounds and is typically carried out between 900 and 1000°C. The char is further reacted with steam in the second gasifier to produce syngas. Lastly, the water gas shift reaction is used to determine the final ratio of CO_2 to H_2 (Fig. 3.3).

3.1.6 Separation of Product Gases

The product syngas of any of the above processes can be separated using the pressure swing adsorption (PSA) process. This process operates by separating different species under pressure according to these species affinity for an adsorbent material. The adsorptive material is used to separate gases, adsorbing the undesired gases at high pressure leaving a stream of hydrogen to pass through. The adsorbent material gradually becomes saturated with the waste gases. The adsorbent bed is depressurized allowing the adsorbed waste gas molecules to flow out of the bed and the process is repeated. The desorbed gas is sent back to the furnace for combustion.

3.1.7 Water Electrolysis

Water electrolysis is a process in which water is broken into hydrogen and oxygen by passing an electric current through the water. This process provides only a

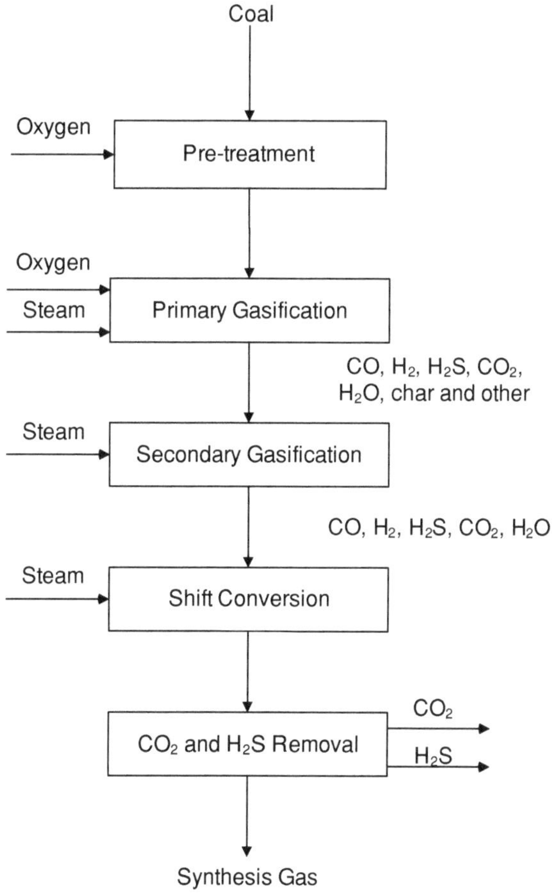

Fig. 3.3 Coal gasification block diagram [3]

small percentage of the world's hydrogen, most of which is produced for applications requiring small volumes of high purity hydrogen. Early electrolysis cells were about 60–75% efficient, but small-scale units now achieve efficiencies closer to 80–85%. Larger units are usually a little less efficient at 75–80% [5].

3.1.7.1 Thermodynamic Analysis

At the cathode of an electrolytic cell, hydrogen ions accept electrons in a reduction reaction that produces hydrogen gas:

$$2H_3O + (aq) + 2e- \rightarrow H_2(g) + 2H_2O(l) \tag{3.4}$$

At the anode, hydrogen ions give up electrons to the anode in an oxidation reaction to form water and oxygen gas:

$$3H_2O(l) \rightarrow {}^1\!/\!_2O_2(g) + 2H_3O + (aq) + 2e- \quad (3.5)$$

The overall reaction is:

$$H_2O(l) \rightarrow H_2(g) + {}^1\!/\!_2\, O_2(g) \quad (3.6)$$

The decomposition of water into hydrogen and oxygen at standard conditions is not thermodynamically favorable. The cathode half-cell potential, determined by using Eq. 3.7, is given below. The potential U° for the cathode half reaction is 0 V, but because $[H_3O^+]$ is not 1 M, but is instead 10^{-7} M, U differs from $\mathring{U}$ and equals:

$$U(\text{cathode}) = 0 - \frac{RT}{nF}\ln\frac{P_{H_2}}{[10^{-7}]^2} \quad (3.7)$$

If the hydrogen is produced at atmospheric pressure, this becomes:

$$\begin{aligned} U(\text{cathode}) &= 0 - \frac{RT}{nF}\ln\frac{1}{[10^{-7}]^2} \\ &= -0.414\,V \end{aligned} \quad (3.8)$$

For the anode half reaction, the half cell potential is:

$$U(\text{anode}) = U^\circ - \frac{RT}{nF}\ln\frac{1}{(P_{O_2})^{1/2}[H_3O^+]^2} \quad (3.9)$$

The standard potential given at standard conditions is $U^\circ = 1.229$ V. Assuming O_2 is at atmospheric pressure:

$$\begin{aligned} U(\text{cathode}) &= 1.229\,V - \frac{RT}{nF}\ln\frac{1}{[10^{-7}]^2} \\ &= 0.815\,V \end{aligned} \quad (3.10)$$

The overall cell voltage is then:

$$\begin{aligned} \Delta U &= U(\text{cathode}) - U(\text{anode}) = -0.414 - 0.815\,V \\ &= -1.229\,V \end{aligned} \quad (3.11)$$

The negative ΔU value indicates that the process will not occur without an applied driving potential. The minimum voltage that will allow the reaction to proceed is called the decomposition potential of water and is equal to 1.229 V. When this potential is applied, hydrogen will be produced at the cathode and oxygen at the anode.

The above example is a purely thermodynamic analysis. Because the concentration of H_3O^+ is so low in pure water, the reaction would occur very slowly. An electrolyte solution is usually used in real-world processes to increase the rate of reaction.

3.1.7.2 Alkaline Water Electrolyzers

In an alkaline electrolyzer, liquid electrolyte systems typically use a corrosive solution such potassium hydroxide. In these systems, oxygen ions migrate through the electrolytic material, leaving hydrogen gas dissolved in the water stream. This hydrogen is extracted from the water and directed into a separating tank.

3.1.7.3 PEM Electrolyzers

In a PEM electrolyzer cell, hydrogen ions are drawn into and through the membrane, where they recombine with electrons to form hydrogen molecules. Oxygen gas remains behind in the water. Water is recirculated and oxygen accumulates in a separation tank.

3.2 Hydrogen Storage

Hydrogen storage is an important component in the ongoing development and commercialization of hydrogen and fuel cell technologies for transportation-related and stationary applications. Transportation applications of hydrogen storage have thus far received the most attention by governments and companies who are trying to commercialize the technology. To make progress in this technology, the US Department of Energy issued a roadmap with milestones to be met by 2010 and 2015, shown in Table 3.1. The performance characteristics of interest for viable hydrogen storage solutions are volumetric (kWh/L) and gravimetric (kWh/kg) energy densities, cost, refueling time, discharge kinetics and cycle life.

In this section, we will cover the various hydrogen storage technologies currently available and some of those that are being researched. These include compression, liquefaction, metal hydride storage, chemical hydrides, carbon-based storage and liquid carrier-based storage.

3.2.1 Compressed Hydrogen Storage

The compression of hydrogen for the purpose of storage is a well-established technology. It is the most simple of existing storage technologies, requiring only a compressor and storage vessel capable of being pressurized. Typical tank materials include steel, aluminum wrapped in fiberglass and high molecular weight lined tanks with a carbon composite shell. Steel tanks can be used where weight and volume are not a constraint on system requirements such as in stationary applications. The two other tank types mentioned are used when weight and volume are significant constraints on system design such as vehicular applications.

Table 3.1 DOE hydrogen storage targets [9]

Storage parameter	Units	2007	2010	2015
DOE technical targets: on-board hydrogen storage systems				
System gravimetric capacity Usable, specific-energy from H_2 (net used energy/max system mass)[a]	kWh/kg (kg H_2/kg system)	1.5 (0.0045)	2 (0.06)	3 (0.009)
System volumetric capacity Usable energy density from H_2 (net used energy/max system volume)	kWh/L (kg H_2/L system)	1.2 (0.036)	1.5 (0.045)	2.7 (0.081)
Storage system Cost[b]	\$/kWh net (\$/kg H_2)	6 (200)	4 (133)	2 (67)
(& fuel cost)[c]	\$/gge at pump	–	2–3	2–3
Durability/operability				
• Operating ambient temperature[d]	°C	−20/50 (sun)	−30/50 (sun)	−40/60 (sun)
• Min/Max delivery temperature	°C	−30/85	−40/85	−40/85
• Cycle life (1/4 tank to full)[e]	Cycles	500	1000	1500
• Cycle life duration[f]	% of mean (min) at%confidence	N/A	90/90	99/90
• Min delivery pressure from tank; FC = fuel cell, I = ICE	Aim (abs)	8FC/10ICE	4FC/35ICE	3FC/35ICE
• Max delivery pressure from tank[g]	Aim (abs)	100	100	100
Charging/discharging rates				
• System fill time (for 5 kg)	min	10	3	2.5
• Minimum full flow rate	(g/s)kW	0.02	0.02	0.02
• Start time to full flow (20°C)[h]	s	15	5	5
• Start time to full flow (−20°C)[h]	s	30	15	15
• Transient response 10%–90% and 90%–0%[i]	s	1.75	0.75	0.75
Fuel Purify (H_2 from strorage)[j]	%H_2	99.99 (dry basis)		

(continued)

Table 3.1 (continued)

Storage parameter	Units	2007	2010	2015
Environmental Health & Safety				
• Penmeation & leakage[k]	Scc/h	Meets or exceeds applicable standards		
• Toxicity	–			
• Safety	–			
• Loss of useable H_2[l]	(g/h)/kg H_2 stored	1	0.1	0.05

[a]Generally the 'full' mass (including hydrogen) is used, for systems that gain weight, the highest mass during discharge is used

[b]2003 US$; total cost includes any component replacement if needed over 15 years or 150,000 mile life

[c]2003 US$; includes off-board costs such as liquefaction, compression, regeneration, etc.; 2015 target based on H_2 production cost of $2 to $3/gasoline galloon equivalent untaxed, independent of production pathway

[d]Stated ambient temperature plus full solar load. No allowable performance degradation from −20C to 40C. Allowable degradation outside these limits is TBD

[e]Equivalent to 100,000; 200,000; and 300,000 miles respectively (current gasoline tank spec)

[f]All targets must be achieved at end of file

[g]In the near term, the forceout should be capable of delivering 10,000 psi compressed hydrogen, liquid hydrogen, or chilled hydrogen (77 k) at 5,000 psi. In the long term, it is anticipated that delivery pressures will be reduced to between 50 and 150 atm for solid storage systems, based on today's knowledge of sodium alanetes

[h] Flow must initiate within 25% of target time

[j]The storage system will not provide any purification, but will receive incoming hydrogen at the purity levels required for the fuel cell. For fuel cell systems, purity meets S AE J2719, Information Report on the Development of a Hydrogen Quality Guideline in Fuel Cell Vehicles. Examples include; total non-particulars,100 ppm; H_2O, 5 ppm; total hydrocarbons (C_1 basis), 2 ppm; O_2, 5 ppm; He, N_2, Ar combined, 100 ppm; CO_2, 1 ppm; CO, 0.2 ppm; total S, 0.004 ppm; formaldehyde (HCHO), 0.01 ppm; formic acid (HCOOH), 0.2 ppm; NH_3 0.1 ppm; total halogenates, 0.05 ppm; maximum particular size, <10 μm, particulate concentration, <1 μg/L H_2. These are subject to change. See Appendix F of DOE Multiyear Research, Development and Demonstration Plan (www.eere.energy.gov/hydrogenandfuelcells/mypp/) to be updated as fuel purity analyses progress. Note the some storage technologies may produce contaminants for which effects are unknown; these will be addressed as more information becomes available

[k]Total hydrogen lost into the environment as H_2; relates to hydrogen accumulation in enclosed spaces. Storage system must comply with CSA/NGV2 standards for vehicular tanks. This includes any coating or enclosure that incorporates the envelope of the storage system

[l]Total hydrogen lost from the storage system, including leaked or vented hydrogen; relates to loss of range

The main limitation with using compression as a means of hydrogen storage is the low energy densities achieved. The gas can be further compressed to a higher storage pressure, but this results in higher capital and operating costs. At 350 bar (~5,075 psi) and 700 bar (~10,150 psi), the energy densities achieved are ~2.6 MJ/L and 4.4 MJ/L, respectively, compared with 31.6 MJ/L for gasoline. Significant increases in volumetric energy density are not achieved at higher pressures.

As the storage pressure is increased, the work of compression increases. The adiabatic compression work required to compress a gas is given in Eq. 3.12 below.

$$W = \left[\frac{\gamma}{\gamma - 1}\right] p_0 V_0 \left[\frac{p_1}{p_0}^{\frac{\gamma-1}{\gamma}} - 1\right] \tag{3.12}$$

where:

W ≡ Work
γ ≡ Ratio of specific heats
p_0 ≡ Initial pressure
p_1 ≡ Final pressure
V_0 ≡ System volume

The isothermal compression work is given by Eq. 3.13.

$$W = p_0 V_0 \ln\left(\frac{p_1}{p_0}\right) \tag{3.13}$$

The actual compression work falls in between these two values. For multistage compression of hydrogen at a rate of 1000 kg/h from 1 bar to 200 bar, the compression work has been determined to be 7.2% of the energy content of hydrogen based on its HHV [6]. Figure 3.4 shows the adiabatic, isothermal and multistage compression work as % of the higher heating value (HHV) of hydrogen (142 MJ/kg) versus pressure assuming initial temperature and pressure as STP values.

Quantum Technologies has recently developed a high pressure hydrogen containers by validating the first 800 bar hydrogen storage tank (TriShield™). The inner shell of this container is made of a high molecular weight polymer to form a

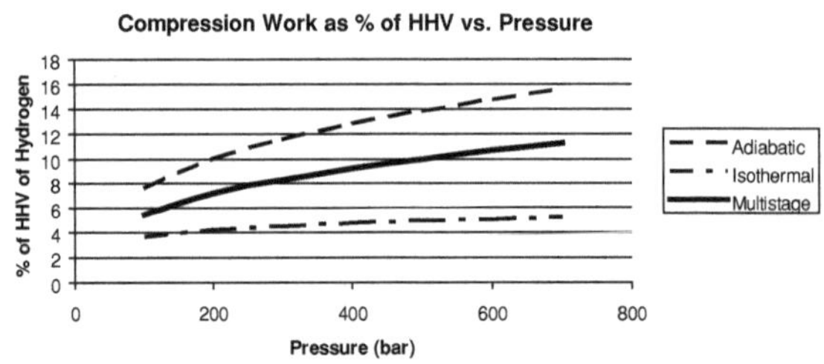

Fig. 3.4 Compression work vs. storage pressure

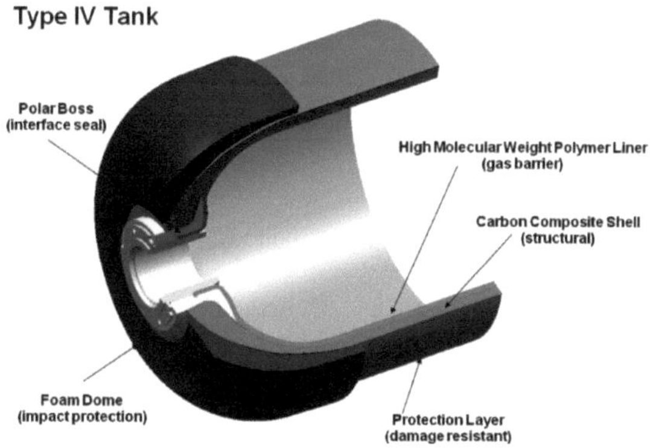

Fig. 3.5 Quantum high pressure hydrogen tank [7]

gas permeation barrier, whereas the outer shell is steel or an aramid-material to prevent the tank from external damage and the structural material in-between is carbon fiber. The carbon fiber represents the largest portion of all costs and so the goal is to reduce the amount of carbon fiber needed while maintaining equivalent levels of performance and safety [6, 7] (Fig. 3.5).

3.2.2 Hydrogen Liquefaction and Storage

Liquid hydrogen is stored in cryogenic tanks at 21.2 K at ambient pressure. Because of the low critical temperature of hydrogen (33 K), the liquid can only be stored in open systems. Liquefaction of hydrogen is done by cooling gaseous hydrogen to form a liquid. The simplest liquefaction process is the Joule-Thompson expansion cycle. In this process, the gas is compressed, cooled in a heat exchanger, and passed through a throttling valve where it undergoes isenthalpic expansion, which results in the production of some liquid.

The liquid is removed and the gas is returned to the compressor via a heat exchanger. The work of liquefaction is ~15.2 kWh/kg, which is almost half the lower heating value of hydrogen.

Once liquefied, the hydrogen must be stored in an insulated vessel. A major concern in liquid hydrogen storage is minimizing hydrogen losses from liquid boil-off. Heat transfer from the environment to the liquid causes some hydrogen to evaporate. The vessel can be refrigerated to prevent boil-off losses, but this consumes energy. If not refrigerated, the hydrogen gas can be vented or captured, liquefied and returned to the tank. These losses are typically 0.4% per day for 50 m^3 double-walled, vacuum insulated tanks [8]. The high energy needed to liquefy hydrogen and the boil-off losses dramatically increase the cost to store hydrogen through this method (Fig. 3.6).

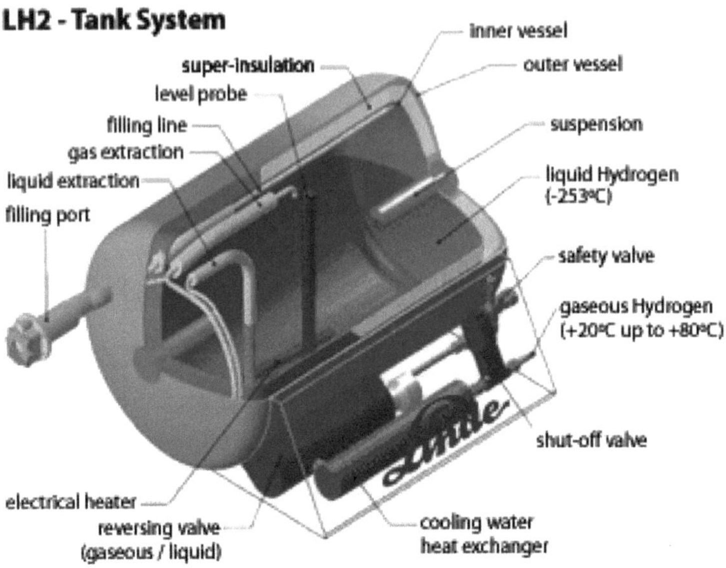

Fig. 3.6 Linde liquid hydrogen storage tank [9]

3.2.3 Metal Hydrides

Metal hydrides store hydrogen by chemically bonding the hydrogen to metal elements and alloys. Heat is released when a hydrogen storage container is filled. When the hydrogen pressure is increased, the hydrogen dissolves in the metal and then begins to bond to the metal. Heat released during hydride formation must be removed to prevent the hydride from heating up. When hydrogen is released, the tank requires heat to be transferred to it at a rate proportional to the rate at which hydrogen is being released. There are many different alloys that can be used and each alloy has different performance characteristics.

Compounds containing hydrogen bonds are known for every metal and non-metal (expect for noble gases) in the periodic table. They can be divided into three groups of hydrides. The first and second group forms a saline whereas the transition metals form mainly metallic compounds. The covalent hydrides can be found at the right of the transition metals. Many of these compounds, MH_n, show large deviations from ideal stoichiometry ($n = 1, 2, 3$). They are also called interstitial hydrides as hydrogen often appears on the interstitial sites in the metal lattice.

The intermetallic phases are of particular interest for hydrogen storage. The properties of these hydrides can be tailored because of the large variation of the elements of the intermetallic compound. The simplest case is the ternary system AB_xH, where A is usually a rare earth or an alkaline metal and B a transition metal.

Table 3.2 Important families of hydride-forming intermetallic compounds [9]

Intermetallic compound	Prototype	Structure
AB_5	$LaNi_5$	Haucke phases, hexagonal
AB_2	ZrV_2, $ZrMn_2$, $TiMn_2$	Laves phase, hexagonal or cubic
AB_3	$CeNi_3$, YFe_3	Hexagonal
A_2B_7	Y_2Ni_7, Th_2Fe_7	Hexagonal
A_6B_{23}	Y_6Fe_{23}	Cubic
AB	TiFe, ZrNi	Cubic
A_2B	Mg_2Ni, Ti_2Ni	Cubic

Additional families of hydride-forming intermetallic compounds are shown in Table 3.2.

3.2.4 Complex Hydrides

Another means of storing hydrogen is via metal-hydrogen complexes. Their light weight and high number of hydrogen atoms per metal atom is an advantage over metal hydrides described previously. The main difference between the metallic and complex hydrides is the transition to an ionic or covalent bond. A large variety of these hydrides are formed by the light metals of group 1, 2 and 3 of the periodic table, i.e. Li, Mg, B and Al. Common examples are $M(BH_4)$ and $M(AlH_4)$ with M = Na or Li as a counter ion.

Using complex metal hydrides for hydrogen storage is a very promising method, since they potentially have a high gravimetric and volumetric hydrogen density. The use of these complex hydrides is challenging because of the thermodynamic and kinetic limitations. Future work involves the improvement of thermodynamic and kinetic properties of existing compounds and exploring new compounds with better practical characteristics for hydrogen storage.

3.2.5 Carbon Based Storage

The physisorption of gas molecules onto the surface of a solid is based on van der Waals forces. As the interaction is weak, high physisorption is observed at low temperatures. There has been extensive research in the late 1990s, particularly on carbon nanotubes. Carbon nanotubes are cylindrical structures with either open ends or closed ends. The closed end variety has a hemispherical end cap at each end. There are two types of carbon nanotubes: multi-walled nanotubes and single-walled nanotubes. MWNTs are composed of multiple, concentric cylindrical tubes.

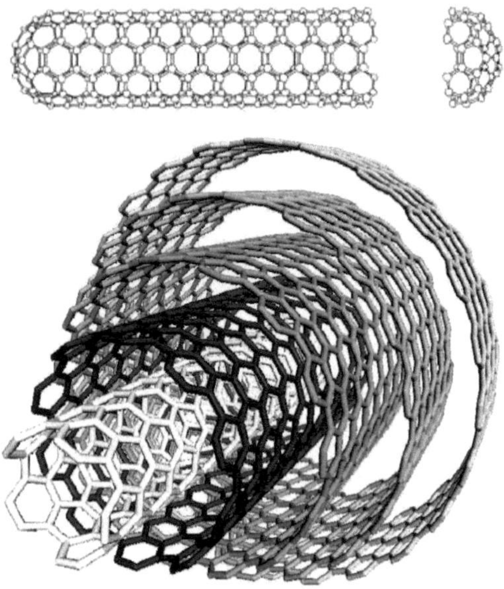

Fig. 3.7 Single-walled (*top*) and multi-walled (*bottom*) carbon nanotube [10, 11]

SWNTs consist of one tube. An example of each is shown in Fig. 3.7. Today the highest gravimetric energy values lay around 4 wt% of hydrogen, but only at low temperatures.

3.2.6 Liquid Carrier Storage

Hydrogen can be "stored" in organic liquids such as ethanol, methanol, butanol and other higher-order C-chain and aromatic ring molecules. The concept is to utilize a reversible hydrogenation/dehydrogenation reaction to store and extract hydrogen from the organic liquid "carrier". The organic liquid can be transported in the same manner as gasoline and the "spent" (dehydrogenated) liquid shipped back for re-hydrogenation. For instance, one mole of naphthalene can be converted to one mole of decalin to provide five moles of H_2. Advantages are that the organic liquids have low volatility, low toxicity and can be made from low-cost raw materials. A major problem of this method however, is the kinetics of both the hydrogenation and dehydrogenation reaction, which often requires a substantial amount of heat and a suitable catalyst for the reaction to occur [12].

3.3 Fuel Cells

Fuel cells are electrochemical systems that convert chemical energy directly into electrical energy. Unlike batteries, in which the required reactants are self-contained, fuel cells require constant streams of fuel and oxidant provided from an external source to provide electricity. The following sections will describe the anatomy of the fuel cell, the functions of the individual components, explain the thermodynamics of fuel cells and describe the various components of complete systems. More detailed information about fuel cells and their applications can be found in the Fuel Cell Handbook published by US Department of Energy [14].

3.3.1 Fuel Cell Structure and Theory of Operation

A fuel cell consists of an electrolyte layer in contact with an electrode on either side. An electrolyte is a material that conducts ions, but prevents the flow of electrons. An electrode is a material that conducts electrons. A schematic representation of a fuel cell constructed with porous electrodes illustrating the reactant/product gases and the ion conduction flow directions through the cell is shown in Fig. 3.8.

The fuel gas enters the cell at the anode (negative electrode) and the oxidant gas enters at the cathode (positive electrode). The electrode is a porous material, so the gas can diffuse through to the opposite side of the electrode where the catalyst is present. The electrode serves several functions. It must capable of transporting the gas stream to the catalyst layer and evenly distribute the gas across the reaction surface area. The electrode must also be capable of diffusing the product gases away from the electrolyte. Lastly, the electrode must have good electrical conductance characteristics because it is responsible for conducting the electrons away from the interface as they are formed in the reaction process.

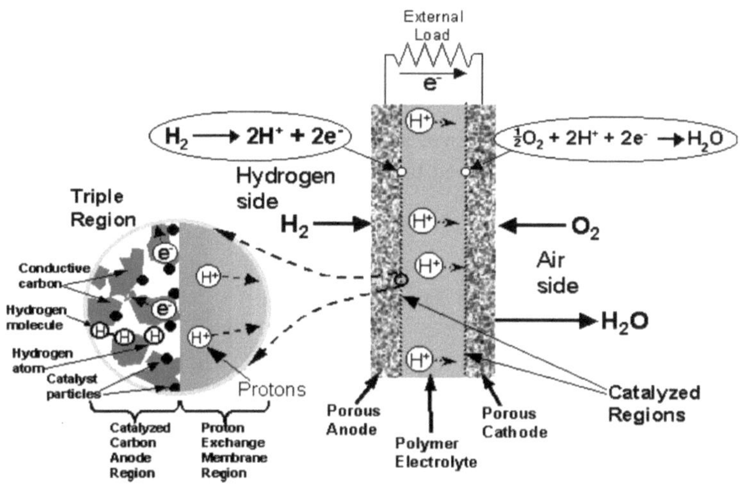

Fig. 3.8 Schematic of a fuel cell [13]

The type of catalyst needed to facilitate proper reaction is a function of temperature. Low temperature fuel cells require expensive catalysts, often platinum, to obtain the desired reaction rates. This condition places a high priority on the designer's ability to reduce the amount of catalyst necessary to provide a given amount of power. At higher temperatures, the reactants can achieve the desired reaction rates using the bulk electrode material.

The electrolyte is also serves several important functions. It must be able to conduct ions from one electrode to the other, but prevent electrons from crossing. It must also act as a barrier between the fuel and oxidant gases to prevent mixing.

In fuel cells with liquid electrolytes, the reactant gases diffuse through an electrolyte film that wets portions of the porous electrode and react on the electrode surface. If the porous electrode contains excess electrolyte, the electrode may block the transport of gaseous reactants to the reaction sites, which results in a reduction in the performance of the fuel cell. Control systems play an important role in maintaining balance among the electrode, electrolyte, and gaseous phases in the porous electrode.

A three-phase interface is established among the reactants, electrolyte, and catalyst in the region of the porous electrode. Depending on the type of fuel cell, molecules at the anode or the cathode will be adsorbed by the catalyst allowing for the formation of ions and electrons. The electrons are conducted away by the electrode to an external circuit which can be used to do work. The charge-carrying ions diffuse through the electrolyte to the opposite electrode and recombine with the electrons. This electrochemical reaction produces both heat and electricity.

In general, the overall reaction in a fuel cell may be written as follows:

$$\text{Fuel} + \text{Oxidant} \rightarrow \text{Oxidation Products} + \text{Useful Work} + \text{Heat Rejected}$$

3.3.2 *Thermodynamics of Fuel Cells*

This section will discuss the thermodynamics of fuel cells. The ideal performance of the fuel cell will first be explained and losses due to the non-ideal conditions of real systems will then be explained to develop equations that describe real-world fuel cell performance.

3.3.2.1 Gibbs Free Energy and Nernst Equation

Total energy is composed of two types of energy: (1) Gibb's free energy, G and (2) unavailable energy, TS. Free energy earns its name because it is energy that is available for conversion into usable work. The unavailable energy is lost due to the increased disorder, or entropy, S, of the system. The maximum electrical work (W_{el}) obtainable in a fuel cell operating at constant temperature and pressure is given by the change in Gibbs free energy (ΔG) of the electrochemical reaction:

$$W_{el} = \Delta G = -nFU \tag{3.14}$$

where:

n ≡ the number of electrons involved in the reaction
F ≡ Faraday's constant (96,487 Coulombs/g-mole electron)
U ≡ the ideal potential or open-circuit voltage of the cell

For a fuel cell, the maximum work available is related to the free energy of reaction, whereas the enthalpy (H) of reaction is the relevant quantity in the case of thermal conversion such as a heat engine. For the state function:

$$\Delta G = \Delta H - T\Delta S \tag{3.15}$$

The difference between ΔG and ΔH is proportional to the change in entropy (ΔS). The maximum amount of electrical energy available is ΔG, as mentioned above, and the total thermal energy available is ΔH. The amount of heat that is produced by a fuel cell operating reversibly is TΔS.

For a general cell reaction:

$$aA + bB \rightarrow cC + dD \tag{3.16}$$

The free energy change is given by the equation:

$$\Delta G = \Delta G^{\circ} + RT \ln\left[\frac{[C]^{c}[D]^{d}}{[A]^{a}[B]^{b}}\right] \tag{3.17}$$

where:

ΔG° ≡ Change is Gibb's free energy at standard pressure (1 atm) and temperature, T
R ≡ Universal gas constant
T ≡ Temperature
[A] ≡ Concentration of species A

Substituting Eq. 3.14 in Eq. 3.17, we get:

$$U = U^{\circ} + \frac{RT}{nF}\ln\left[\frac{[A]^{a}[B]^{b}}{[C]^{c}[D]^{d}}\right] \tag{3.18}$$

This is the general form of the Nernst equation. It relates the ideal potential to the concentrations of reactants and products.

3.3.3 *Cell Efficiency and Polarization*

The thermal efficiency of an energy conversion device is defined as the amount of useful energy produced relative to the change in stored chemical energy. The ideal efficiency of a fuel cell, operating reversibly, is given by:

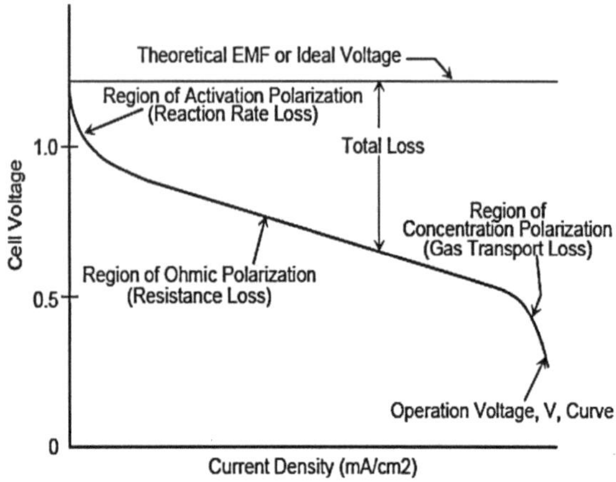

Fig. 3.9 Cell voltage vs. current density [14]

$$\eta = \frac{\Delta G}{\Delta H} \tag{3.19}$$

The performance of a real-world fuel cell is less than that given in Eq. 3.19 because of losses due to (a) activation polarization, (b) ohmic polarization and (c) concentration polarization. *Polarization* refers to the departure of the potential from equilibrium conditions due to the flow of current. *Overpotential* refers to the magnitude of this departure. The plot shown in Fig. 3.9 shows the typical behavior of cell potential as a function of current density (defined as current drawn per unit surface area of the fuel cell).

The horizontal line represents the ideal cell voltage obtained from the Nernst equation. The actual cell voltage begins to decrease as current starts flowing in the circuit.

3.3.3.1 Activation Polarization

In order to begin the flow of current in the fuel cell, a potential must be applied to overcome reduce the activation energy of the rate limiting step of the chemical reaction taking place. This potential is called the activation overpotential and is denoted η_s. The rate of reaction can be related to the activation overpotential by the Bulter-Volmer equation, shown as Eq. 3.20:

$$i = i_0 \left[\exp\left(\frac{\alpha_a F}{RT} \eta_s \right) - \exp\left(\frac{\alpha_c F}{RT} \eta_s \right) \right] \tag{3.20}$$

A positive value of η_s produces an anodic current, which means that the electrons will be stripped from thc reactants transferred to the electrode. A negative value will lead to a cathodic current, which means that electrons will be transferred from the electrode to the reactants. The variable i_0 is called the exchange current density.

This value is a function of the concentration of reactants and products, temperature and the electrode/electrolyte interface. A high exchange current density value suggests that good fuel cell performance is possible because high reaction rates are possible with smaller activation overpotential. α_a and α_c are called transfer coefficients and are parameters that relate how the applied potential favors the direction of the reaction occurring.

3.3.3.2 Ohmic Polarization

As the current increases, the ohmic resistance to the flow also increases. This is known as *ohmic polarization*. This polarization loss increases nearly linear with the current. The ohmic polarization is described by the following equation.

$$\eta_{ohm} = IR \tag{3.21}$$

where:

I ≡ Total current flowing through system
R ≡ Total resistance of the cell

3.3.3.3 Concentration Polarization

As the current density increases, diffusion resistance effects begin to play an important role because the fuel and the oxidant do not transport quickly enough to the electrode surface. This is known as *concentration polarization*.

Under short-circuit conditions, the system potential reaches zero and the limiting current, i_L, is reached. The limiting current is the maximum rate at which the reactant can be supplied to an electrode. The concentration polarization is given by the following equation.

$$\eta_c = \frac{RT}{nF} \ln\left(1 - \frac{i}{i_L}\right) \tag{3.22}$$

3.3.3.4 Overall Cell Potential

The cell potential will depend on the open-circuit potential, activation overpotential, ohmic potential drop and concentration overpotential. Therefore, the actual cell voltage is given as:

$$V = U + \eta_{a,anode} - \eta_{a,cathode} - \eta_{ohm} + \eta_{c,anode} - \eta_{c,cathode} \tag{3.23}$$

3.3.4 Fuel Cell Types

Fuel cells are usually named based on the type of electrolyte used in that particular fuel cell. The theory underlying the operation of fuel cells has been given in the previous sections. This section will summarize the different types of fuel cells and differences among them.

3.3.4.1 Polymer Electrolyte Membrane Fuel Cell

The polymer electrolyte membrane fuel cell (PEFC) utilizes a hydrated polymeric membrane that acts as an ion conductor. Nafion is a widely used membrane manufactured by DuPont. The electrodes are porous carbon electrodes with a platinum catalyst. The most common fuel is pure H_2 and the most common oxidant is air. The hydrogen enters at the anode, where it is ionized with the help of a platinum catalyst. The hydrogen atoms are conducted across the membrane while the free electrons are conducted in an external circuit to the cathode where they meet with the hydrogen and air to form water. The platinum is highly sensitive to carbon monoxide, so the use of fuels other than pure H_2 is challenging and costly. Much research has been aimed at decreasing the amount of platinum that is used to reduce the price of the fuel cell.

Water management is a critical issue because a lack of water will decrease the cell life and ionic conductivity, while excess water will decrease the power output. In order to maintain proper hydration and maximize power, the cell operating temperature is held around 80°C.

Besides water management, temperature and pressure also play important roles in determining fuel cell performance. Increasing the temperature of PEM fuel cell is advantageous because of the reduction of ohmic resistance of the electrolyte at higher temperatures and because of reduced mass transfer limitations. The poisoning of the catalyst by CO is also reduced at higher operating temperatures.

Another important performance variable is pressure. Higher oxygen pressure reduces cathode polarization, which improves the fuel cell efficiency. Another advantage is that higher pressure decreases the membrane dehydration at a given temperature, so higher temperatures are possible. The performance improvement at high pressure must be balanced against the membrane stability and energy required in compressing oxygen. Operating pressures of 2–3 atmospheres lead to an optimal balance between cost and performance.

3.3.4.2 Phosphoric Acid Fuel Cells (PAFC)

Phosphoric acid is used for the electrolyte in this fuel cell. Typical operating temperatures range from 150 to 220°C. At higher temperatures, phosphoric acid is a good ionic conductor and, as in the case of PEFC's, CO poisoning of the

Pt is reduced. In addition, the use of concentrated acid minimizes allows for much simpler water management.

The PAFC can operate at a range of temperatures, pressures, gas utilization rates and current densities. The benefit of finding the most effective operating conditions must be balanced against the energy required to achieve these conditions in a given environment as well as the effect on the system life. An increase in the operating temperature increases cell voltage, but this increase must be limited to avoid catalyst and component wear. Utilization of a higher percentage of the fuel is desirable from a cost standpoint, but results in lowered partial pressure, which leads to higher concentration polarization losses. Lastly, cell voltage, and thus cell efficiency, decreases with an increase in current density.

3.3.4.3 Molten Carbonate Fuel Cells

Molten Carbonate fuel cells have (MCFC) great potential to emerge as the most efficient type of fuel cell. They have the attractive combination of being able to work at high efficiencies with natural gas fuel, while not being hampered by the same corrosion concerns as solid oxide fuel cells.

The electrolyte in this fuel cell is a combination of alkali carbonates. The fuel cell operates in the range of 600–700°C. At these higher operating temperatures, the nickel and nickel oxide electrode material are sufficient to catalyze the reaction. Also, due to the high operating temperatures, MCFCs generate a lot of waste heat that can easily be used in cogeneration applications. The disadvantages of operating at such high temperatures are increased corrosion and reduced cell component life.

3.3.4.4 Solid Oxide Fuel Cells

Solid oxide fuel cells (SOFC) operate at very high temperatures, which make them ideal for cogeneration applications. The high temperature also allows them to operate on a variety of fuels. Units have been developed that operate on methane, propane, butane, gasified biomass, and paint fumes. SOFC technology is still in the early stages of development compared to the other fuel cell types, so the potential for rapid advancement is present.

The electrolyte in an SOFC is usually yttria-stabilized-zirconia (YSZ). The operating temperature ranges from 600 to 1000°C. The anode is most often made of a Co-YSZ or Ni-YSZ cermet and the cathode is Sr-doped $LaMnO_3$.

This high temperature fuel cell has several attractive characteristics with regard to central power generation. In addition, a very high temperature allows internal reforming of various fuels.

Table 3.3 Summary of fuel cell technologies [14]

	PEM	PAFC	MCFC	SOFC
Fuel Processing	Contaminant Removal; Reforming; LT Shift; HT Shift	Contaminant Removal; Reforming; HT Shift	Contaminant Removal	Possible Contaminant Removal
Operating Temp., °C	80	200	650	1000
Peak Power Density, mW/cm^2	~700	~200	~160	~150–200 (tubular)
System Electrical Efficiency, % HHV	32–40	36–45	43–55	43–55
Waste Heat Temp., °C	~60–80°C	~50–90°C	150–350°C	~600–800°C
Start-up Time, h	<0.1	1–4	5–10	5–10
kW electricity vs. kW heat	47%/53%	44%/56%	50%/50%	60%/40%
NO_x (g/MWh)	<20	<10	<10	<10
SO_x (g/MWh)	<0.1	<0.1	<0.1	<0.1
Performance degradation, %/1000 h	>1	0.44	0.60	<0.10
Water Consumption	0	90	88	0
Current Applications	Transportation Residential- Cogeneration	Limited Cogeneration	Central Power Cogeneration	Central Power Cogeneration
Costs ($/kW installed)[a]	5,000	4,500	2,800	3,500
Fuel Costs ($/kWh)[b]	0.4	0.076	0.060	0.064

[a]Based on 200 kW system

[b]For natural gas price of $8/MMBtu, or for PEM H_2 price of $40/MMBtu

3.3.5 *Summary Table of Fuel Cell Characteristics*

A brief summary of comparison for these four major types of fuel cells is shown in Table 3.3.

References

1. Bharadwaj SS, Schmidt LD (1995) Catalytic partial oxidation of natural gas to syngas. Fuel Process Technol 42(2–3):109–127
2. Reyes SC, Sinfelt JH, Feeley JS (2003) Evolution of processes for synthesis gas production: recent developments in an old technology. Ind Eng Chem Res 42(8):1588–1597
3. Lee S (1997) Methane and its derivatives, 1st edn. Marcel Dekker, Inc, New York
4. Armor JN, Martenak DJ (2001) Studying carbon formation At elevated pressure. Appl Catal A Gen 206:231–236
5. Towards a Hydrogen Economy. Research Reports International (2004)
6. Bossel U, Baldur E (2003) Energy and the hydrogen economy. 10–11
7. Andrighetti J (2006) Quantum hydrogen storage systems. Hydrogen Storage Challenges for Mobility, 7 Dec 2006, SAE International
8. Brunner TA (2006) Liquid Hydrogen Storage- Roadmap to Mass Market. Hydrogen Storage Challenges for Mobility, 7 Dec 2006, SAE International
9. HFCIT Hydrogen Storage: Gaseous and Liquid Hydrogen Storage. US Department of 2007 http://www1.eere.energy.gov/hydrogenandfuelcells/storage/hydrogen_storage.html
10. Digital image [Carbon nanotubes]. University of Karlsruhe Germany. 7 June 2007. http://www.ipc.uni-karlsruhe.de/mik/download/nt1.jpg
11. Digital image [Mutli-walled carbon nanotube]. Université Libre de Bruxelles. 7 June 2007 http://www.ulb.ac.be/sciences/cpmct/MWNT_oblique.gif
12. More A (1996) Modern production technology—ammonia, methanol, hydrogen, carbon monoxide—a review. CRU, London
13. Digital image [PEM Mechanism]. Third Orbit Power Systems. 7 June 2007 http://www.thirdorbitpower.com/images/PEMmechanism.GIF
14. Fuel Cell Handbook. U.S. Department of Energy. 2004

Chapter 4
Operation and Control of Hybrid Energy Systems

This chapter will describe the operation of a Renewable Hybrid Energy System (RHES). Long-term simulations used for system sizing will also be discussed.

4.1 Renewable Hybrid Energy Systems: System Configuration and Theory of Operation

Renewable hybrid energy systems can reduce the cost of high-availability renewable energy systems. This results from the system's ability to take advantage of the complementary diurnal (night/day) and seasonal characteristics of available renewable resources at a given site. Energy storage costs can be reduced in the same way as a result of the complementary characteristics of the storage devices used.

Hybrid power systems are designed for the generation and use of electrical power. They may consist of an AC distribution system, a DC distribution system, multiple electricity generation that utilize several energy resources, energy storage, rectifiers, inverters, user loads, heating and dispatchable loads and a system controller.

An example hybrid power system shown in Fig. 4.1 operates by using electricity produced from renewable energy sources to meet an electrical demand load. In the system under consideration, the renewable energy sources are wind and solar radiation. A wind turbine converts wind energy into electrical energy and photovoltaic (PV) panels convert solar energy, in the form of photons, into electrical energy. This energy is then directed to the primary load, a battery bank, electrolyzer, heating load or dispatchable load. A dispatchable load is one that can be met when energy is available, which differs from a primary load that must be met immediately when demanded. In systems that are grid connected, the excess energy could be sold to the electrical grid. The system could be operated so that energy is sold to the grid at times when the utility is paying higher rates to electricity producers.

S. Al-Hallaj and K. Kiszynski, *Hybrid Hydrogen Systems*,
Green Energy and Technology, DOI: 10.1007/978-1-84628-467-0_4,

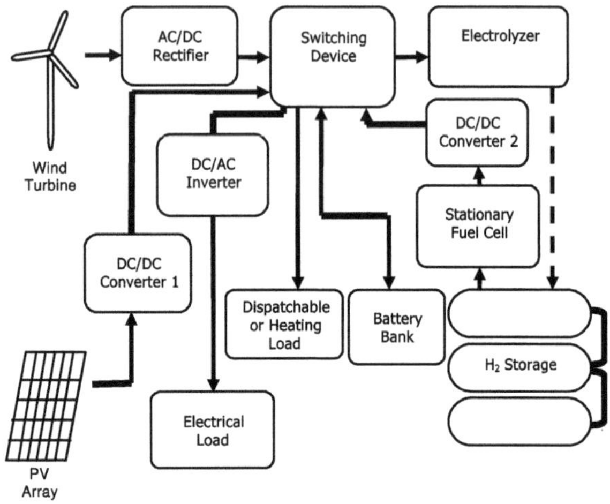

Fig. 4.1 Schematic of stand-alone hybrid system

4.1.1 Factors Influencing RHES

There are several factors influencing how an RHES is sized and how it is operated. They are:

Energy Demand
Available Renewable Energy Resources
System Component Capability and Costs

Energy demand will depend on user habits, the specific application and seasonal variations. Available renewable energy resources are determined by: (1) location, defined by latitude, longitude and altitude; (2) time, both short term (daily fluctuations) and long term (over the course of the year); and (3) surrounding land topology. Lastly, system component specifications and costs determine how the various components can be sized and how they can be operated. Initial component costs, maintenance costs and component lifetime will affect which combination of components and what operating scheme should be chosen to minimize the system lifetime costs. It is because of the large number of variables and the interaction among the many system components that simulation software is often necessary to design a hybrid energy system.

4.1.2 System Simulation for Component Sizing Purposes and Determination of System Cost

The simulation software used for sizing the RHES was written in MATLAB (Natick, Massachusetts) and is based on freely available software known as HOMER (NREL, Golden, CO). It uses hourly average wind speed, incident solar

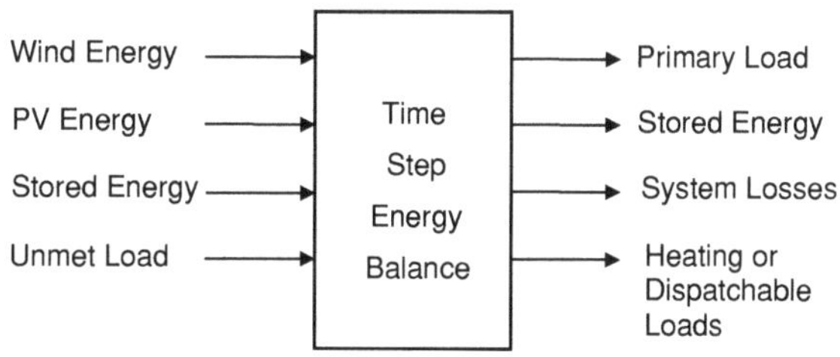

Fig. 4.2 Schematic of energy balance

radiation and user demand data for an entire year to carry out the system simulations. Hourly data is used for several reasons. First, finding data on the order of minutes or seconds is very difficult. Data available to carry out these simulations is almost always averaged over the hour. Second, carrying out a simulation using high frequency data over an entire year would significantly increase simulation time. Lastly, hourly data has been shown to be sufficient to determine long term performance of systems and perform economic analyses of these systems [1–3]. The software simulates the system using various combinations of component sizes and searches for the system that meets the user constraints at the lowest cost.

4.1.3 Explanation of Energy System Simulation

The time series simulation employs an energy balance approach within each time step. Energy is therefore conserved throughout the entire simulation. The schematic, shown in Fig. 4.2, shows the energy sources and sinks for the simulation being performed.

At the beginning of the simulation for a given system configuration, the initial state of the system is set. The program reads the incident solar radiation data, wind speed data and demand data for the given hour from the user provided data files. The amount of electrical energy that can be produced for the given system is calculated and compared to the demand. If excess energy is present, the energy is stored. If there is insufficient energy, the stored energy is used to make up the difference. Storage of excess energy and utilization of stored energy are described in Sects. 4.6 and 4.7.

4.2 Storage of Excess Energy

When multiple means of storage are used, the dispatch strategy can have a significant impact on the cost of operation and reliability of the system [4–10]. The dispatch strategy describes the logic that governs which energy stores are used under certain circumstances. In these systems, when there is a shortage of energy from the primary renewable resources to meet the energy demand, decisions need to be made to determine how best to utilize the stored energy. In case where both a battery bank and hydrogen storage are used, several scenarios may arise.

Use the battery bank to meet the demand. Can the battery bank provide the power demanded? What is the minimum state of charge of the battery bank? How long will the battery bank be able to meet the demand load at the required discharge rate? How does discharging the battery bank at such a discharge rate affect battery cycle life and thus the system operating costs?

Use the fuel cell to meet the demand load. Does the fuel cell have the required capacity to meet the demand load? At what operating point does the electrolyzer operate most efficiently? Should it be operated to produce more power than what is required so that the battery bank can be recharged?

Use both the battery and fuel cell simultaneously to meet the demand load. How should the battery and fuel cell share the load? All of the questions mentioned above need to be addressed when controlling the systems and can lead to complex dispatch strategies to optimize system performance. They are highly dependent on component cost, dynamics, degradation mechanisms and other component characteristics.

4.2.1 Battery Constraints

There are several constraints imposed on the variable $Dec(1)$ that defines the fraction of excess energy that the battery bank can accept during charging. The first constraint is the maximum charge power as determined by the kinetic battery model for the entire battery bank. Different battery chemistries will affect the choice of model used and constraints. The kinetic battery model is commonly used for lead acid batteries [11]. The maximum battery charge power is given by

$$Pbatt_{\mathrm{cmax,kbm}} = \frac{V_{\mathrm{nom}} I_{\mathrm{c,max}} N_{\mathrm{batt}}}{1,000} \tag{4.1}$$

where:

N_{batt} = the number of batteries in the battery
bank $I_{\mathrm{c,max}}$ = the battery's maximum charge current as determined by the kinetic battery model (A)
V_{nom} = the battery's nominal voltage (V)

The second constraint relates to the maximum charge rate of the battery, which is the charge current (A) per Ah of unused battery capacity. The battery charge power in kW corresponding to this maximum charge rate is given by the following equation:

$$Pbatt_{\mathrm{mcr}} = \frac{N_{\mathrm{batt}}\left(1 - e^{-\alpha_c \Delta t}\right)\left(q_{\max} - q_0\right)}{1,000} \tag{4.2}$$

where:

α_c = battery's maximum charge rate (A/Ah)

The third constraint relates to the battery's maximum charge current specified by the manufacturer. The maximum battery bank charge power corresponding to this maximum charge current is given by the following equation:

$$Pbatt_{mcc} = \frac{N_{\text{batt}} I_{\max} V_{\text{nom}}}{1{,}000} \quad (4.3)$$

where:

N_{batt} = the number of batteries in the battery bank
$I_{\max}$ = the battery's manufacturer specified maximum charge current (A)
V_{nom} = the battery's nominal voltage (V)

The smallest of these three values is chosen as the charge power constraint, Eq. 4.4.

$$Pbatt_{\text{cmax}} = \min\left(P_{bott,kbm}, P_{bott,mcr}, P_{bott,mcc}\right) \quad (4.4)$$

This value is divided by the excess energy for the current hour, as shown in Eq. 4.5, to determine the fraction of excess energy that can be sent to the batteries.

$$Pbatt_{rloadc} = \frac{Pbatt_{\text{cmax}}}{D_1} \quad (4.5)$$

where:
D_1 = the amount of excess energy available during the current hour
Finally, the constraint that is actually applied to determine the fraction of the excess energy that can be stored in the battery for the current hour is determined as shown in Eq. 4.6. The smallest of these values will be chosen.

$$Dec1_{con} = \min(P_{batt,rload}, 1) \quad (4.6)$$

As a result of the constraints imposed, the battery bank will never be charged too quickly or charged beyond its full capacity. The last remaining constraint, unity, prevents the optimizer from considering charging the battery with more energy than that which is available.

4.2.2 Electrolyzer Constraints

Two inequality constraints are applied to *Dec*(2), the fraction of excess energy that can be supplied to the electrolyzer during a given hour. The first sets the lower operating limit of the electrolyzer. This value, divided by the amount of excess power available, determines the lower constraint. Please see the following equation

$$Dec2L_{Con} = \frac{I_{\min} Elec_{Size}}{D_1} \quad (4.7)$$

where:

$Elec_{Size}$ = the rate capacity of the electrolyzer
$I_{\min}$ = fraction of rated capacity at which the electrolyzer can operate

The second and third electrolyzer constraints are determined by Eqs. 4.8 and 4.9. They ensure that the maximum rated power input of the electrolyzer is not exceeded and also ensure that the amount of power sent to the electrolyzer does not exceed the power available.

$$Perc_{rated,elec} = \frac{Elec_{Size}}{D_1} \tag{4.8}$$

$$Dec2H_{Con} = \min(Perc_{rated,elec}, 1) \tag{4.9}$$

4.2.3 Hydrogen Storage Constraints

In the case of hydrogen storage, a constraint is imposed to prevent the hydrogen tank from being overfilled. A MATLAB function, *fzero*, searches the interval from 0 to 1 inclusive to determine the value $Dec2_{Tank}$ in Eq. 4.10, if it exists. This is the value that would fill the tank completely.

$$\left(\frac{(100 - Tank_{Level}) \cdot H2_{TankSize}}{100}\right) + \left(D_1 \cdot Dec2_{Tank} \cdot Elec_{polyfun}(Dec(2), D_1, Elec_{Size}) \cdot \left(\frac{3.6}{120}\right)\right) = 0 \tag{4.10}$$

where:

$Tank_{Level}$= the level of the hydrogen tank (%)
$H2_{TankSize}$= size of the hydrogen tank (kg)
D_1 = excess energy (kWh)
$Dec2_{Tank}$ = upper constraint for $Dec(2)$ that prevents the hydrogen tank from being overfilled (unitless)
$Elec_{Size}$ = size of the electrolyzer (kW)

$Elec_{polyfun}$ is a function shown in Eq. 4.11 that accepts the argument $\left(\frac{Dec(2) \cdot D_1}{Elec_{Size}}\right)$. It is the equation for the electrolyzer efficiency curve.

$$Elec_{polyfun}(x) = 0.7485x^2 - 1.8308x + 1.4777 \tag{4.11}$$

If the value determined in this search is smaller than the value of $Dec2H_{Con}$ found in Sect. 4.3.4, then the old value of $Dec2H_{Con}$ is replaced with the new value. Otherwise, it remains unchanged.

4.2.4 Energy Storage Logic

If there is excess energy, the *Excess_standard* program is called to determine how to store the excess energy. In this case, $Dec(1)$ is the variable that specifies the

Table 4.1 Conditions and resultant actions for energy storage

Scenario	Conditions	Action
I	$Dec1_{Con} = 1$	$Dec(1) = 1$ $Dec(2) = 0$
II	$Dec1_{Con} < 1$ $((1-Dec1_{Con})*D_1)/Elec_{Size} < I_{min}$	$Dec(1) = Dec1_{Con}$ $Dec(2) = 0$
III	$Dec1_{Con} < 1$ $((1-Dec1_{Con})*D_1)/Elec_{Size} \geq I_{min}$ $1-Dec1_{Con} \leq Dec2H_{Con}$	$Dec(1) = Dec1_{Con}$ $Dec(2) = 1-Dec1_{Con}$
IV	$Dec1_{Con} < 1$ $((1-Dec1_{Con})*D_1)/Elec_{Size} \geq I_{min}$ $1-Dec(1) > Dec2H_{Con}$	$Dec(1) = Dec1_{Con}$ $Dec(2) = Dec2H_{Con}$

fraction of the excess energy that is sent to the battery bank and *Dec*(2) is the variable that specifies the fraction of energy that is sent to the electrolyzer. There are four possible scenarios. Please see Table 4.1 for the conditions that must be satisfied for different actions to be taken. The different scenarios are evaluated in the order shown in the table. As a result, at least one condition at each of the previous scenarios is not satisfied. Scenario I applies if the battery can accept all of the energy. Scenario II applies if the battery cannot accept all of the energy, but the amount that cannot be stored by the battery is insufficient to operate the electrolyzer. Scenario III applies when the battery cannot accept all of the energy, there is enough energy to operate the electrolyzer above its minimum operating point and, lastly, there is tank capacity to store the hydrogen produced by the electrolyzer. Finally, Scenario IV applies when the battery cannot accept all of the energy, there is enough energy to operate the electrolyzer above its minimum operating point, but the tank does not have enough capacity to store all of the hydrogen that could be produced by the electrolyzer.

4.3 Utilization of Stored Energy

In the case where there is not enough energy available, the "Supplement Optimizer" program is called to determine how energy should be routed from the fuel cell and battery bank. Again, the optimization problem consists of a cost function and several constraints.

4.3.1 Battery Constraints

When the battery is discharged, two constraints are applied. The first constraint, determined by Eq. 4.12, ensures that the battery is not discharged at a rate higher than that determined by the kinetic battery model. The second constraint, determined by Eq. 4.13, prevents the battery from being depleted beyond the minimum state of charge specified by the battery manufacturer. The least of these two values and unity is chosen as shown in Eqs. 4.14–4.16. The value of unity prevents the optimization algorithm from taking more energy from the battery than would be needed to meet the entire demand load.

$$Pbatt_{dmax,kbm} = \frac{\left(N_{batt} \cdot V_{nom} \cdot I_{d,max}\right)}{1000} \tag{4.12}$$

where:

V_{nom} = nominal battery voltage (V)
$I_{d,\max}$ = maximum discharge current (A)

$$Perc_{dis,batt} = \eta_{battD} \cdot \frac{(SOC_{batt} - (100 - DOD_{max}))}{100} \cdot \frac{V_{nom} C_{nom} N_{batt}}{1000} \tag{4.13}$$

where:

η_{battD} = discharge efficiency of battery
SOC_{batt} = battery state of charge (%)
C_{nom} = nominal storage capacity (Ah)

$$Pbatt_{dmax} = \min\left[Pbatt_{d\,max,kbm}, Perc_{dis,batt}\right] \tag{4.14}$$

$$Pbatt_{rloadd} = \frac{Pbatt_{dmax}}{-D_1} \tag{4.15}$$

$$Dec1_{Con} = min\left[Pbatt_{rloadd}, 1\right] \tag{4.16}$$

4.3.2 Fuel Cell Constraints

Three constraints are applied to fuel cell operation. The first prevents the fuel cell from meeting a load larger than that for which it is rated and is determined by Eq. 4.17. The second prevents the fuel cell from providing more power than that which can flow through the DC/DC converter that is connected to the output of the fuel cell. This constraint is determined by Eq. 4.18. The least of these two values and unity is chosen as the constraint that is actually applied during optimization. The third constraint prevents the fuel cell from providing more power than is required to completely meet the demand load, as shown in Eq. 4.19:

$$Perc_{rated,FC} = \frac{FC_{Size}}{-D_1} \tag{4.17}$$

where:

FC_{Size} = the rated capacity of the fuel cell (kW)

$$FC_{Conv2} = \frac{Conv2_{Size}}{-D_1} \tag{4.18}$$

where:

$Conv2_{Size}$ = the rated capacity of converter #2

$$Dec2_{Con} = \min(Perc_{rated,FC}FC_{Conv2}) \tag{4.19}$$

4.3.3 Hydrogen Utilization Constraint

A constraint is applied to the hydrogen tank so that it is not depleted beyond empty. When the fuel cell is to be used, the system logic verifies that the hydrogen tank is above 5% of the full tank capacity. This is an arbitrary value, but is used because the fuel cell is unlikely use more than 5% of the tank capacity in any given time step. If the optimization is carried out, a constraint is applied to prevent the hydrogen tank from dropping below zero, as shown by Eq. 4.20. It accepts the argument$\left(\frac{Dec(2)*(-D_1)}{FC_{Size}}\right)$. Equation 4.21 is the function fit to the fuel cell efficiency curve [7].

$$\begin{aligned} Perc_{cap,FC} = -\left(\frac{Tank_{Level}}{100}\right) \cdot H2_{TankSize} - D_1 \cdot Dec(2) \\ \cdot FC_{poly,fun}(Dec(2), D_1, FC_{Size}) \cdot \frac{3.6}{120} \leq 0 \end{aligned} \tag{4.20}$$

$$FC_{poly,fun}(x) = \frac{-0.07491x^3 - 0.04543x^2 + 0.5844x + 6.633 \times 10^{-5}}{x + 6.696 \times 10^{-3}} \tag{4.21}$$

4.3.4 Energy Utilization Logic

Two different sets of energy management logic (EML) have been developed to determine how to utilize stored energy. They are described in the next several sections.

4.3.4.1 EML: Standard

If there is an energy deficit and the "standard" control logic is chosen *Supplement_Standard* is called. The battery bank is used to meet as much of the demand

Table 4.2 Conditions and resultant actions using "Standard" utilization control logic

Scenario	Conditions	Action
I	$Dec1_{Con} \geq 1$	$Dec(1) = 1$ $Dec(2) = 0$
II	$Dec1_{Con} < 1$ $Tank_{Level} \geq 5$ $Dec2_{Con} \geq 1 - Dec1_{Con}$	$Dec(1) = Dec1_{Con}$ $Dec(2) = 1 - Dec1_{Con}$
III	$Dec1_{Con} < 1$ $Tank_{Level} \geq 5$ $Dec2_{Con} < 1 - Dec1_{Con}$	$Dec(1) = Dec1_{Con}$ $Dec(2) = Dec2_{Con}$
IV	$Dec1_{Con} < 1$ $Tank_{Level} < 5$	$Dec(1) = Dec1_{Con}$ $Dec(2) = 0$

load as possible and the fuel cell is used as backup if the battery bank cannot meet the full demand load. Table 4.2 shows the conditions and corresponding actions taken.

4.3.4.2 EML: Optimal

If there is an energy deficit and "optimal" control logic is chosen, the *Supplement_Optimizer* program is called to determine the optimal way to provide the energy required for the demand load that the wind turbines and solar array cannot. As above, $Dec(1)$ is the variable that specifies the fraction of energy that comes from the batteries and $Dec(2)$ specifies the fraction of energy that comes from the fuel cell. Table 4.3 shows the conditions and corresponding actions taken.

When there is not enough energy to meet the demand load by the primary renewable energy sources, four scenarios must be considered. If there is enough stored energy considering both of the hydrogen tank and battery bank, the hydrogen tank is above 5% of its full capacity, and both constraints, $Dec1_{Con}$ and $Dec2_{Con}$ are greater than zero, then the optimization can be carried out. The cost function for removing energy from the energy stores is given by Eq. 4.22 and is to be minimized.

$$\begin{aligned} S_1 = C_{E,batt} \cdot (-D_1) \cdot Dec(1) + \frac{N_{batt} \cdot C_{batt} \cdot \eta_{batt,D} \cdot (-D_1) \cdot Dec(1)}{2Q_{lifetime,ave}} \\ + C_{H_2} \cdot FC_{poly,fun}(Dec(2), D_1, FC_{Size}) \cdot (-D_1) \cdot Dec(2) \cdot \frac{3.6}{120} \end{aligned} \tag{4.22}$$

where:

$C_{E,batt}$ = cost of energy stored in battery (USD)
C_{H_2}= cost of hydrogen stored in tank (USD)

The first term is the cost of the energy that is discharged from the battery bank. The second term is the wear cost of discharging the battery bank. The last term is

Table 4.3 Conditions and resultant actions using "Optimal" utilization control logic

Scenario	Conditions	Action
I	$Dec1_{Con} + Dec2_{Con} \geq 1$ $Tank_{Level} > 5$ $Dec1_{Con} > 0$ $Dec2_{Con} > 0$	*Optimize Eq.* 4.22
II	$Tank_{Level} > 5$ $Dec2_{Con} > 0$	$Dec(1) = Dec1_{Con}$ $Dec(2) = Dec2_{Con}$
III	None of the above	$Dec(1) = Dec1_{Con}$ $Dec(2) = 0$

the cost of the hydrogen used. The cost of the fuel cell wear is not included in the cost function because it is a fixed cost and is not affected by the decision variables.

Two inequality constraints are applied to *Dec*(1). The constraint on the left hand side is a linear constraint. The constraint on the right hand side is classified as a non-negativity constraint.

$$Dec1_{Con} \geq Dec(1) \geq 0 \tag{4.23}$$

Three inequality constraints are applied to *Dec*(2). The constraint on the left hand side is a linear constraint. The constraint on the right hand side is classified as a non-negativity constraint. Equation 4.24 is the third constraint and is non-linear.

$$Dec2_{Con} \geq Dec(2) \geq 0 \tag{4.24}$$

The last constraint is an equality constraint and ensures that the amount of energy required by the demand load is provided.

$$Dec(1) + Dec(2) = 1 \tag{4.25}$$

If the first set of conditions cannot be satisfied, a second scenario must be considered. In the second scenario, if the hydrogen tank is more than 5% full and $Dec2_{Con}$ is greater than zero, then $Dec(1) = Dec1_{Con}$ and $Dec(2) = Dec2_{Con}$. This is done because if the first set of conditions were not met, then there was not enough energy to fully meet the demand load. By setting the decision variables equal to their respective upper constraints, the maximum possible fraction of the demand load will be met.

The third scenario to be considered occurs when the first two sets of criteria cannot be met. In this case, if $Dec1_{Con}$ is greater than zero, then $Dec(1) = Dec1_{Con}$ and $Dec(2) = 0$. This is done because if the previous sets of criteria have not been met, then the fuel cell cannot meet any of the demand load. This is due to either the hydrogen tank containing too little hydrogen or there being no fuel cell in the system.

The fourth and last scenario occurs when the battery is at its minimum state of charge and the fuel cell cannot meet any portion of the load. In this case, both decision variables are set equal to zero. None of the demand load can be met.

If the demand load cannot be completely met, the cumulative demand load that has not been met is tracked by a variable named *Unmet_load*. This variable is checked later in the main program to determine if the system availability is below the user specified requirement. If it is the case, the program aborts simulation of the current system and moves on to the next system. This is done to reduce simulation time. The less capable a system is to meet the demand load, the sooner it will fail to meet the user's desired availability and the sooner simulation of the system will be terminated.

4.4 System Update

The state of the system is now updated. If there was an excess of energy, the battery bank, electrolyzer and hydrogen tank information is updated. If there was an insufficient amount of energy, the battery bank, fuel cell and hydrogen tank information is updated.

4.4.1 Excess Energy: System Update

The battery subsection updates the following:

1. The state of charge of the battery.
2. The value of energy stored in the battery bank.
3. The amount of energy that may be cycled through the battery bank through the rest of its life.
4. The available charge in the battery bank based on the kinetic battery model.
5. The bound charge in the battery bank.

The state of charge of the battery is updated using the following equation.

$$SOC_{batt} = SOC_{batt,0} + \frac{100D_1 \cdot Dec(1) \cdot \eta_{battC}}{\frac{V_{nom}C_{nom}N_{batt}}{1000}} \tag{4.26}$$

where:

SOC_{batt} = battery state of charge at end of time period (%)
$SOC_{batt,0}$ = battery state of charge at beginning of time period (%)

The value of the energy stored in the battery bank is updated using the following equation.

$$C_{E,batt} = \frac{C_{E,batt,0} \cdot SOC_{Batt} + \left(\frac{C_{Turbine,kWh} \cdot E_{Wind} + C_{PV,kWh} \cdot E_{PV}}{\eta_{battC}(E_{Wind} + E_{PV})}\right) \cdot Change_{Batt,SOC}}{SOC_{Batt} + Change_{Batt,SOC}} \tag{4.27}$$

where:

$C_{E,batt,0}$ = cost of energy stored in the battery bank at beginning of time period (USD/kWh)
$C_{Turbine,kWh}$ = cost of energy produced by the wind turbine (USD/kWh)
E_{Wind} = energy produced by wind turbine during time period (kWh)
$C_{PV,kWh}$ = cost of energy produced by the PV array (USD/kWh)
E_{PV} = energy produced by PV array during time period (kWh)
$Change_{Batt,SOC}$ = change in the battery state of charge (%)

The amount of energy that may be cycled through the battery bank during the rest of its life is updated using the following equation.

$$Life_{batt} = Life_{batt,0} - \frac{D_1 \cdot Dec(1)}{2} \tag{4.28}$$

where:

$Life_{batt}$ = battery life left at end of time period (kWh)
$Life_{batt,0}$ = battery life left at beginning of time period (kWh)

The values of the available and bound energy are updated using Eqs. 4.26 and 4.27. Lastly, the amount of energy used for other purposes is updated using Eq. 4.29.

$$Total_{Dump,kWh} = Total_{Dump,kWh,0} + D_1(1 - (Dec(1) + Dec(2))) \tag{4.29}$$

where:

$Total_{Dump,kWh}$ = cumulative energy used for other purposes over the year at the end of time period (kWh)
$Total_{Dump,kWh,0}$ = cumulative energy used for other purposes over the year at the beginning of the time period (kWh)

If the electrolyzer was operated during the current hour, the subsection updates:

1. The fraction of the tank that is full.
2. The remaining life of the electrolyzer stack.
3. The value of the hydrogen in the hydrogen tank.

The percentage of the tank that is full is updated using Eqs. 4.30, 4.31 and 4.32.

$$Change_{Tank} = Dec(2) \cdot Elec_{poly,fun}(Dec(2), D_1, Elec_{Size}) \cdot D_1 \left(\frac{3.6}{120}\right) \tag{4.30}$$

$$Change_{Tank,SOC} = \frac{100 \cdot Change_{Tank}}{H2_{Tank,Size}} \tag{4.31}$$

$$Tank_{Level} = Tank_{Level,0} + Change_{Tank,SOC} \tag{4.32}$$

The value of the hydrogen in the tank is updated using Eq. 4.33.

$$C_{H_2} = \frac{C_{H_2,0} \cdot Tank_{Level,0} + \left(\frac{C_{Turbine,kWh} \cdot E_{Wind} + C_{PV,kWh} \cdot E_{PV}}{Elec_{poly,fun}(Dec(2), D_1, Elec_{Size})(E_{Wind} + E_{PV})} \right) \cdot Change_{Tank,SOC}}{Tank_{Level,0} + Change_{Tank,SOC}} \tag{4.33}$$

where:

$C_{H_2,0}$ = cost of hydrogen in storage tank at beginning of time period (USD/kWh)

The value of the electrolyzer stack is updated using the following equation.

$$Life_{elec} = Life_{elec,0} - 1 \tag{4.34}$$

where:

$Life_{elec}$ = electrolyzer life left at end of time period (h)
$Life_{elec,0}$ = electrolyzer life left at beginning of time period (h)

4.4.2 *Energy Deficit: System Update*

The battery subsection updates the same values as in the “Excess Optimizer” section using the same equations except that the sign of the value of D_1 is changed.

If the fuel cell was used during the current hour, the following values are updated:

1. The fraction of the tank that is full.
2. The remaining life of the fuel cell stack.

The percentage of the tank that is full is updated using Eqs. 4.30, 4.31 and 4.32. The remaining life of the fuel cell stack is updated using Eq. 4.35.

$$Life_{FC} = Life_{FC,0} - 1 \tag{4.35}$$

where:

$Life_{FC}$ = fuel cell life left at end of time period (h)
$Life_{FC,0}$ = fuel cell life left at beginning of time period (h)

4.5 Financial Analysis and Sorting of Systems

If a system meets the desired availability and the tank is at least as full as it was at the beginning of the year, then financial analysis will be performed to determine which system provides the lowest overall cost over the project lifetime. The results of this analysis are stored in an array called Feasible. After all systems have been simulated and all systems meeting the desired availability constraint are stored in the *Feasible* array, the rows are sorted based upon net present cost.

To determine the net present cost (NPC), which will be used to determine which system costs the least, several values must be calculated. The initial cost of capital (ICC) is the sum of the cost of all of the system components at the beginning of the project. The annualized cost of capital (ACC) is then determined using the following equation.

$$ACC = ICC \cdot \frac{i(1+i)^{R_{proj}}}{(1+i)^{R_{proj}} - 1} \tag{4.36}$$

where:

i = the interest rate
R_{proj} = the project lifetime

The ACC is the value of annual cash flows for each year of the project discounted to the present time.

The annualized replacement cost (ARC) of each component is calculated based on the projected component lifetime and replacement cost using Eqs. 4.37 and 4.38. It is used to determine the annualized replacement cost of the component minus the salvage value at the end of the project.

$$ARC = C_{rep} \cdot f_{rep} \frac{i}{(1+i)^{R_{comp}} - 1} - C_{rep} \frac{R_{rem}}{R_{comp}} \frac{i}{(1+i)^{R_{proj}}} \tag{4.37}$$

where:

C_{rep} = replacement cost of each component
R_{rep} = replacement cost duration of component
R_{comp} = component life
R_{rem} = remaining life of component at end of project life

$$f_{rep} = \left\{ \begin{matrix} \frac{(1+i)^{R_{proj}}((i+1)^{R_{rep}} - 1)}{(1+i)^{R_{rep}}((i+1)^{R_{proj}} - 1)}, & R_{rep} > 0 \\ 0, & R_{rep} = 0 \end{matrix} \right\} \tag{4.38}$$

The ACC and the ARC of each component is then added to determine the total annualized cost of the system including both the initial capital cost and the annualized replacement costs. By using Eq. 4.39, the NPC is then calculated.

$$NPC = \frac{C_{atot}((1+i)^{R_{proj}} - 1)}{i(1+i)^{R_{proj}}} \tag{4.39}$$

where:

C_{atot} = sum of ACC and each component's ARC

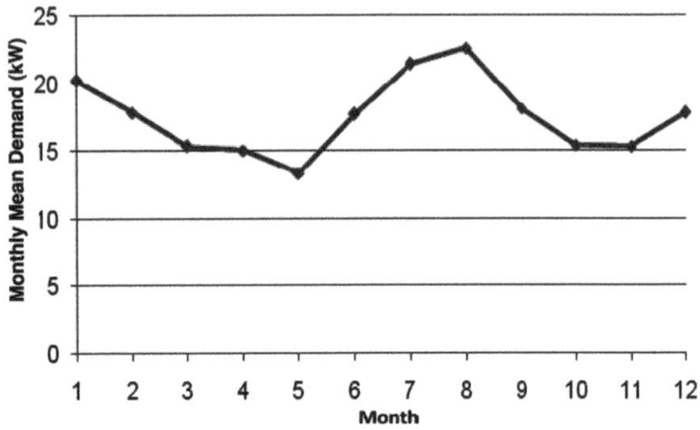

Fig. 4.3 Monthly demand load data for the Rockford Airport fire station in Rockford, IL

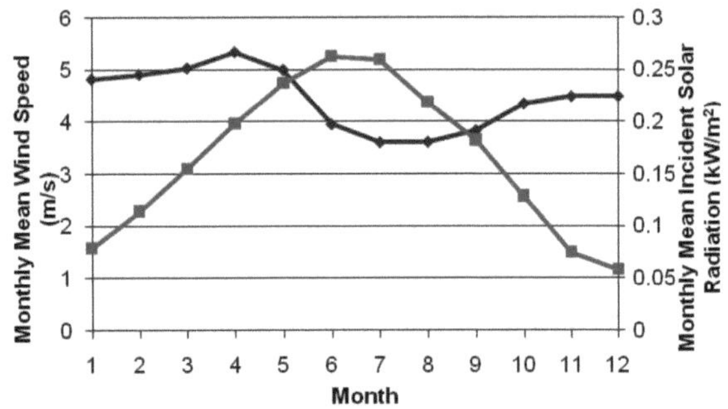

Fig. 4.4 Monthly mean wind speed and mean incident solar radiation for Rockford, IL

4.6 Simulation of Hybrid Energy Systems for a Case study in Rockford, Illinois

4.6.1 Wind alone versus Hybrid wind-solar system

To help develop an understanding of how various components of these energy systems affect each other and overall system performance, a simplified system consisting of a wind farm and battery bank will be studied. The monthly demand load data, shown in Fig. 4.3, and monthly wind speed and solar radiation data, shown in Fig. 4.4, are based on data collected for the Rockford Airport fire station in Rockford, IL.

The wind turbine and battery bank capacities necessary to meet the demand load at the lowest cost will depend on the battery performance characteristics and cost, wind turbine performance characteristics and cost, the temporal mismatch between wind energy availability and energy demand, and required system availability.

To illustrate the economic benefits of the hybridization strategy, a system using only wind turbines is compared to a system that uses both wind turbines and a photovoltaic array. The components considered and relevant specifications are

Table 4.4 Hybrid system component specifications and economic data

Component	Cost/Unit capacity	Replacement cost	Lifetime (Years)
Wind turbine	1.96 USD/W	1 USD/W	25
PV panel	4.73 USD/W	4.73 USD/W	25
Battery	0.42 USD/Ah	0.42 USD/Ah	12
Inverter	1 USD/W	1 USD/W	15

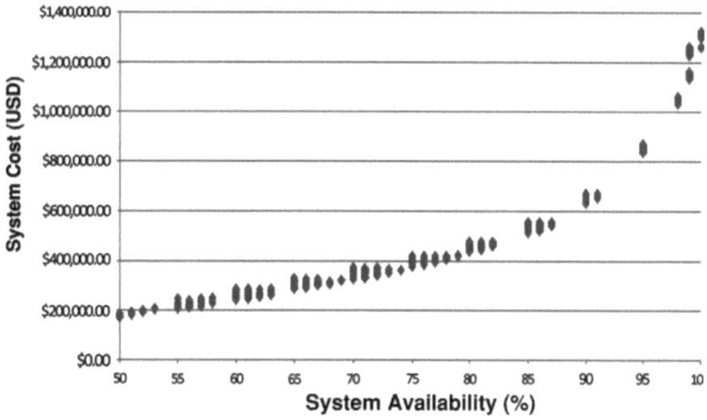

Fig. 4.5 System cost versus system availability

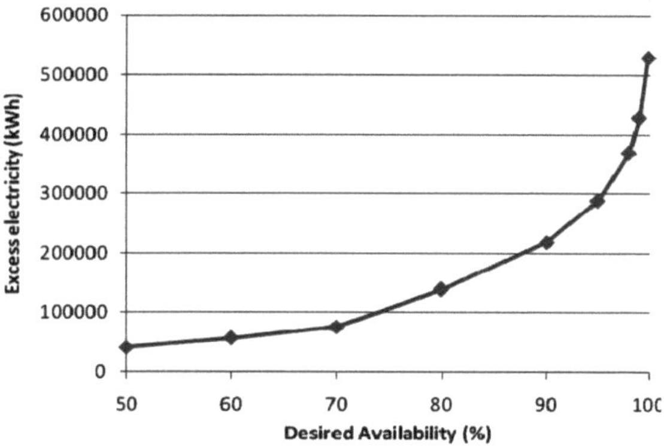

Fig. 4.6 Excess energy versus desired availability

described with economic data in Table 4.4. The user demand in this case is for the fire station for the Rockford Airport in Rockford, IL. A plot of system cost versus desired availability is shown in Fig. 4.5 and the excess electricity produced plotted against desired availability is shown in Fig. 4.6.

The solar panels were modeled as fixed-mount, south facing panels at a 42.35° between panel and horizontal. The wind turbine was modeled after a commercially available WES18 wind turbine using a hub height of 31 m. The variation of wind speed with height was modeled as a logarithmic profile with roughness factor of 0.15 m.

As the desired availability increases, the cost increases non-linearly. Additionally, the amount of excess energy produced increases. Why? From a cost perspective, it does not make sense to increase the battery bank capacity further to

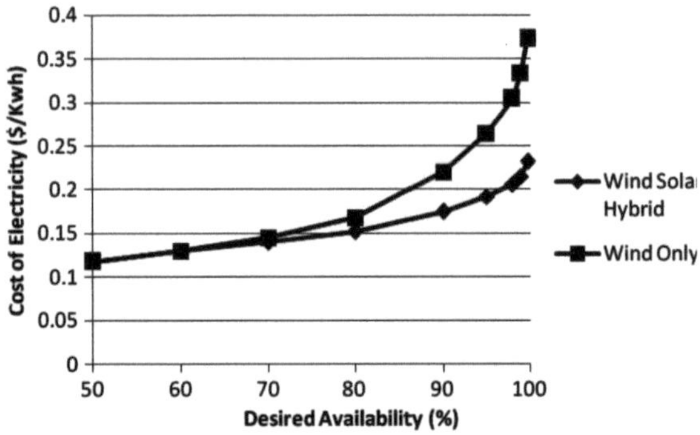

Fig. 4.7 Cost of electricity versus desired availability for wind only and wind-solar hybrid systems

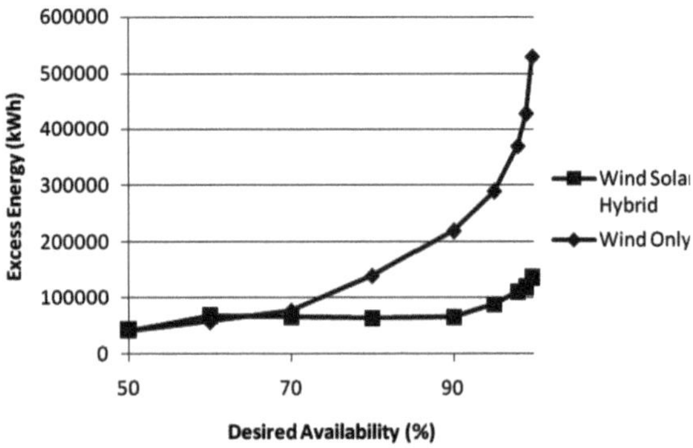

Fig. 4.8 Excess electricity versus desired availability for wind only and wind-solar hybrid systems

Table 4.5 Performance comparison of wind only and hybrid systems

	Rated turbine capacity (kW)	PV array (kWp)	Battery storage capacity (Ah)	Excess production (kWh)	Net present cost (USD)
Wind only	415	0	674,500	529,000	0.373
Wind-PV	110	80	494,000	134,000	0.232

store this energy. It makes more economic sense to oversize the wind turbine and dump this energy than to store it. In situations such as this, using this energy to meet deferrable loads, i.e. loads that can be met at any time, or heating loads may be investigated.

Figures 4.7 and 4.8 show cost of electricity and amount of excess electricity, respectively, plotted against desired availability for wind only and wind-PV hybrid systems.

Economic benefits, based on NPC and cost of electricity, were obtained in using the optimal control techniques presented in this chapter. For a system with 99.8% availability, the cost of electricity is reduced by 37.8% and excess energy beyond what is needed to meet the primary load is reduced by 74.7%. Table 4.5 provides a

Table 4.6 Component information for the hybrid hydrogen system

Component	Name	Cost/Unit capacity	Replacement cost	Lifetime
Wind turbine	WES18	1.96 USD/W	1 USD/W	25 years
Battery	Surrette 4KS25P	0.42 USD/Ah	0.42 USD/Ah	10,569 kWh or 12 years
Inverter	None	1 USD/W	1 USD/W	15 years
Electrolyzer	HOGEN 40	5.5 USD/W	3 USD/W	4000 h
H2 Tank	None	100 USD/kg	100 USD/kg	25 years
Fuel Cell	HyPM 12	4 USD/W	3.25 USD/W	4380 h

Table 4.7 Other simulation parameters

Parameter	Value
Desired availability	99.8%
Project lifetime	25 years
Interest	6%

summary of the performance comparison of wind only and hybrid wind-PV systems.

4.6.2 Hybrid Hydrogen System

Several simulations were run to determine if there is any benefit to using the control strategy devised compared to the standard control logic using the Rockford, IL wind and solar data and the Rockford fire station electricity demand data. Table 4.6 lists the component information used in the simulations.

Table 4.7 lists the other parameters used in the simulations. These include: the system's desired availability, which is defined as the met demand load divided by the total demand load; the project lifetime and the interest rate, which is used in calculating system costs. The tank and battery SOC initial conditions were varied to determine their effect on system cost.

The notation used in this chapter, B40T50Opt for instance, indicates that the system was simulated with the battery SOC set initially to 40% and tank pressure set initially to 50% of the maximum level using the optimization logic described in the previous chapter. B70T50Stan, as another example, indicates that the system was simulated with the battery SOC initially set to 70% and tank pressure set to 50% using the standard logic.

Table 4.8 shows the economic results of the simulations using both the standard and newly devised control logic for different tank pressure and battery SOC initial conditions. These include: the initial capital cost (ICC), which is the initial cost of the system; the annualized capital cost (ACC), which is the capital outlay for each year assuming the interest rate in Table 4.7; the annualized replacement cost (ARC), which is the cost of replacing the components each year; the net present

Table 4.8 Financial results of hybrid hydrogen system simulations

		ICC (USD)	ACC (USD)	ARC (USD)	NPC (USD)	COE (USD)
B40T25	Standard	\$926,428	\$72,471	\$79,023	\$1,010,185	0.371
	Optimal	\$950,770	\$74,376	\$81,668	\$1,043,992	0.353
B40T50	Standard	\$922,450	\$72,160	\$79,609	\$1,017,669	0.389
	Optimal	\$961,484	\$75,214	\$81,921	\$1,047,222	0.354
B40T100	Standard	\$1,015,338	\$79,427	\$87,704	\$1,121,156	0.508
	Optimal	\$1,000,306	\$78,251	\$88,800	\$1,135,159	0.552
B70T25	Standard	\$918,336	\$71,838	\$79,029	\$1,010,255	0.373
	Optimal	\$968,732	\$75,781	\$82,456	\$1,054,060	0.355
B70T50	Standard	\$928,064	\$72,599	\$79,608	\$1,017,659	0.378
	Optimal	\$958,914	\$75,013	\$81,962	\$1,047,746	0.370

Table 4.9 Hybrid hydrogen component sizes for lowest cost systems

		Turbine (kW)	Battery bank (#)	Fuel cell (kW)	Electrolyzer (kW)	Hydrogen tank (kg)
B40T25	Standard	244	129	24	22	575
	Optimal	235	140	25	25	702
B40T50	Standard	225	160	24	20	770
	Optimal	232	121	25	25	1020
B40T100	Standard	299	205	20	17	212
	Optimal	313	270	16	5	87
B70T25	Standard	228	150	24	23	585
	Optimal	236	119	25	25	1025
B70T50	Standard	222	145	24	23	840
	Optimal	247	131	25	20	895

cost (NPC), which is the total cost of the system, including the ACC and ARC discounted to current dollars; and lastly, the cost of electricity (COE), which is the cost per kWh of energy produced by the system.

The NPC for systems using standard logic is less than the NPC for systems using the optimized logic regardless of initial conditions. However, in all cases except B40T100, the COE is less when using the optimized logic. This occurs because the lowest cost systems sized while using the optimized logic favor larger electrolyzers and more frequent use of these electrolyzers, which then increases the load met by the system. Since the load met by the system appears in the denominator of the COE calculation, the COE is reduced. The numerator of the calculation also increases as the costs associated with electrolyzer replacement increase because of the increased number of hours used, but to a lesser extent than the denominator increase.

Table 4.9 lists the system component sizes for the systems simulated. The optimal systems tend to favor larger hydrogen tanks and larger electrolyzers, while the standard systems tend to favor larger battery banks. The one exception is in the case of B40T100. In this case, the turbine and battery capacity are much larger and

Table 4.10 Electrolyzer performance metrics

		Electrolyzer efficiency (%)	Hours of electrolyzer operation	H_2 produced (kg)	Annual electric load (kWh)
B40T25	Standard	40.5	2789	727.4	60170.7
	Optimal	41.0	3220	954.2	78150.1
B40T50	Standard	40.5	2644	626.9	51824.8
	Optimal	41.1	3232	957.8	78336.6
B40T100	Standard	40.3	1180	237.8	19775.7
	Optimal	39.8	1592	94.4	7912.9
B70T25	Standard	40.8	2623	715.2	58882.0
	Optimal	41.0	3279	971.7	79556.0
B70T50	Standard	40.7	2560	698.0	57513.3
	Optimal	40.4	3494	828.5	68675.7

the electrolyzer and hydrogen tank much smaller than systems with lower tank initial conditions. This is due to the fact that the end of the year battery SOC and tank pressure are required to be equal to or greater than the initial conditions. Since the tank needs to be full at the end of the year, a small tank and electrolyzer is used and a large wind turbine capacity is installed to achieve this. Fuel cell size does not appear to depend on initial conditions or on which control logic is used except in the case of B40T100. Wind turbine power generating capacity does not vary considerably except in the case where the hydrogen tank is required to be 100% full at the beginning and end of the year, B40T100.

Table 4.10 lists electrolyzer performance metrics values for the systems. The systems under study behave nearly identically except in the case of the electrolyzer. To identify the reason of the differences in behavior, the systems were simulated and the component sizes were varied to determine the effect of the variation on the electrolyzer operation. The increased number of hours of electrolyzer operation is due to the larger wind turbines. The larger wind turbine in the optimal system increases the number of hours during the year in which the battery cannot accept all of the charge and thus sends the remaining energy to the electrolyzer.

By using the optimal logic, the electrolyzer efficiency is only negligibly higher than in the standard cases. B40T100 and B70T50 are the exceptions. The optimization problem was formulated such that the cost per kWh of energy stored is minimized. Maximization of the amount of energy stored is partially a function of the percentage of the electrolyzer capacity utilized and the efficiency at that capacity. The fact that electrolyzer efficiency in the simulation using optimal logic is nearly that of the electrolyzer run at full capacity, 39.4%, indicates that running the electrolyzer at full capacity minimizes the cost of energy stored as hydrogen most of the time. In other words, the improved efficiency gained by running the electrolyzer at less than full capacity, as can be observed in the electrolyzer efficiency curve, does not compensate for the reduction in hydrogen produced. Under other circumstances, this may not be the case. For instance, there may be

Table 4.11 Fuel cell performance metrics

		Fuel cell efficiency (%)	Hours of fuel cell operation	No. of fuel cell starts	Average operational time (h)	H_2 consumed (kg)	Annual electric load met (kWh)
B40T25	Standard	51.3	750	334	2.2	662.9	11111.3
	Optimal	51.8	996	468	2.1	854.7	14448.3
B40T50	Standard	51.3	712	348	2.0	626.7	10497.4
	Optimal	51.7	1125	538	2.1	957.8	16189.9
B40T100	Standard	49.9	263	106	2.5	237.7	3870.7
	Optimal	48.2	110	44	2.5	94.4	1495.0
B70T25	Standard	51.2	735	350	2.1	647.4	10842.4
	Optimal	51.8	1129	540	2.1	955.5	16163.9
B70T50	Standard	51.3	796	387	2.1	698.0	11700.3
	Optimal	51.7	982	458	2.1	827.2	13988.7

Table 4.12 Battery information

		Battery life (years)	Annual battery throughput (kWh)
B40T25	Standard	12	41701.8
	Optimal	12	39089.6
B40T50	Standard	12	44003.9
	Optimal	12	37649.8
B40T100	Standard	12	45172.9
	Optimal	12	46754.5
B70T25	Standard	12	43334.5
	Optimal	12	37290.3
B70T50	Standard	12	43072.6
	Optimal	12	38617.4

some circumstances under which the demand load and resources are such that the situations where the electrolyzer may run at less than full capacity to achieve higher efficiencies occur frequently enough that over-sizing the electrolyzer capacity to take advantage of the improved efficiency reduces energy storage costs.

Table 4.11 lists fuel cell performance metrics values. Efficiency is slightly higher in all cases where optimal control is used except for the B40T100 system. The number of hours of fuel cell operation is higher in the optimal cases because of the smaller battery bank and hydrogen consumed is higher in the optimal cases because of this increased frequency of operation. The fuel cells meet a larger amount of the annual load in all cases except the B40T100 case.

Table 4.12 lists the battery metrics. The battery throughput is lower in almost all cases where optimal logic is used as this logic favors use of the fuel cell as mentioned above. The battery life in all cases is equal to the float life. The optimal control logic as formulated attempted to reduce costs by reducing the battery throughput. This should reduce replacement cost and, thus, NPC. The battery bank life is limited by either battery throughput or the float life, which, in this case, is 12 years. If the float life is less than that imposed by the battery bank throughput,

Table 4.13 Wind energy production and utilization information

		Hours of turbine operations	Energy produced (kWh)	Unmet load (kWh)	Excess energy (kWh)
B40T25	Standard	6758	397715.2	287.3	180226.4
	Optimal	6758	383045.4	295.5	151558.4
B40T50	Standard	6758	366745.6	256.8	158218.3
	Optimal	6758	378155.5	265.6	148378.5
B40T100	Standard	6758	487364.2	287.9	299439.7
	Optimal	6758	510183.9	304.4	330837.2
B70T25	Standard	6758	371635.6	304.7	156242.6
	Optimal	6758	371635.6	304.7	156242.6
B70T50	Standard	6758	361855.7	271.7	149048.9
	Optimal	6758	402605.2	243.3	179270.6

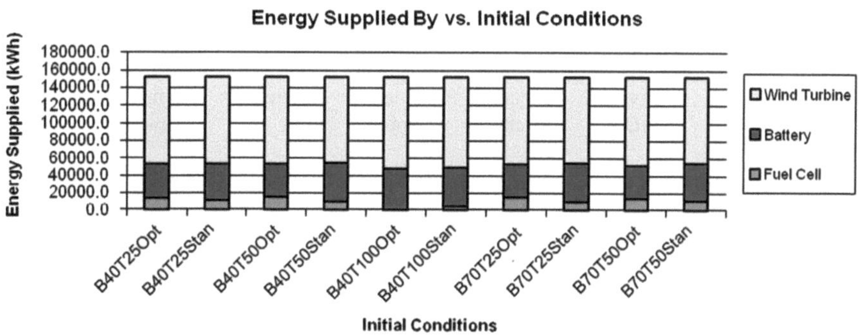

Fig. 4.9 Energy supplies used to meet demand load for standard and optimized control logic

then reducing battery throughput provides no economic benefit as float life is independent of how the battery bank is operated.

Table 4.13 lists wind turbine and energy production metrics. The hours of operation is the same in every case because the same wind data was used for all simulations. The data indicates that as long as the initial conditions of the battery bank SOC and tank pressure remain relatively low, the amount of energy produced in excess is reduced. As the tank or battery initial conditions increase, the amount of excess energy produced increases. Additionally, because the optimal control method favors the use of hydrogen, there is an even greater increase in excess energy produced because the turbine has to be oversized to meet the tank requirements.

Figure 4.9 shows the amount of energy used to meet the demand load from the turbine, fuel cell and battery bank.

Figure 4.10 shows how the energy is used for the systems simulated. By reducing the tank and battery initial conditions, the amount of energy produced in excess is reduced.

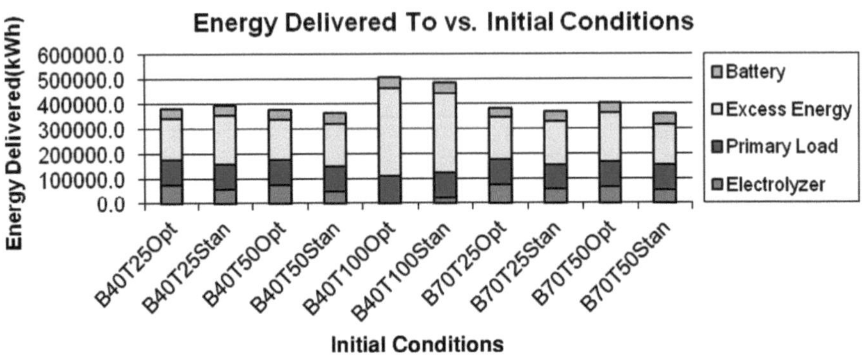

Fig. 4.10 Energy sinks and amount of energy going to each sink

Table 4.14 Mean NPC difference for optimal and standard systems using different tank and battery initial conditions

		No. of unique system ID's	No. of matching unique system ID's	Percent of unique Opt systems w/Lower NPC	Mean NPC difference	Percent of standard systems w/Lower NPC	Mean NPC difference
B40T25	Optimal	1339	102	8.8	$51,593	91.2	$85,140
	Standard	947					
B40T50	Optimal	1178	125	0	NA	100	$109,856
	Standard	723					
B40T100	Optimal	1387	82	1.2	$96,660	98.8	$34,435
	Standard	1059					
B70T25	Optimal	1029	103	42.7	$81,584	57.3	$209,627
	Standard	797					
B70T50	Optimal	632	125	20.8	$53,531	79.2	$50,078
	Standard	732					

The data presented in Table 4.14 is the result of data mining of all of the non-optimal (based on component size) systems simulated. The data was searched so that only systems with unique sizing combinations for a given set of initial conditions would be examined. For examples, this resulted in 1339 unique systems simulated for the B40T25Opt case. This was repeated for optimal and standard control for all sets of initial conditions. Systems using different control techniques, but the same initial conditions and component sizes were then matched based on their system ID's. The NPC for systems with different control schemes for each set of initial conditions were then compared to determine how the systems performed disregarding the component sizing. The results suggest that optimal systems perform better than standard systems as the battery SOC initial conditions increase, but do not perform as well when the hydrogen tank initial conditions increase. Based on the data shown, the magnitude of the performance benefit of one control technique over another as a function of initial conditions cannot be commented on.

4.7 Conclusions

In this chapter, a hybrid energy system simulation program was written in MATLAB. The hybrid system consisted of a wind turbine, lead-acid battery bank, electrolyzer, fuel cell, hydrogen tank and power conversion/conditioning equipment. The simulator defines a component search space and performs an annual simulation for each component combination. It subsequently identifies all feasible systems, sorts them by net present cost and identifies the least cost system. A case study was presented for a wind only compared to a hybrid solar-wind system with battery storage. The demand load and energy resource curves were based on the Rockford Airport fire station energy demand and the wind speed data for that location. Commercially available components were identified and these components' performance and cost specifications were used in the simulations.

References

1. Ai B, Yang H, Shen H, Liao X (2003) Computer-aided design of PV/wind hybrid system. Renew Energy 28:1491–1512
2. Dutton AG, Bleijs JA, Dienhart H, Falchetta M, Hug W, Prischich D, Ruddell AJ (2000) Experiences in the design, sizing, economics, and implementation of autonomous wind-powered hydrogen production systems. Int J Hydrogen Energy 25:705–722
3. Hollmuller P, Joubert J, Lachal B, Yvon K (2000) Evaluation of a 5 kWp photovoltaic hydrogen production and storage installation for a residential home in Switzerland. Int J Hydrogen Energy 25:97–109
4. Kelouwani S, Agbossou K, Chahine R (2005) Model for energy conversion in renewable energy system with hydrogen storage. J Power Sources 140:392–399
5. Muselli M, Notton G, Louche A (1999) Design of hybrid-photovoltaic power generator, with optimization of energy management. Sol Energy 65:143–157
6. Paynter RH, Lipman NH, Foster JE (1991) The potential of hydrogen and electricity production from wind energy. Energy Research Unit, Rutherford Appleton Laboratory, September 1991
7. Santarelli M, Pellegrino D (2005) Mathematical optimization of a RES-H2 plant using a black box algorithm. Renew Energy 30:493–510
8. Seeling-Hochmuth GC (1997) A combined optimization concept for the design and operation strategy of hybrid-PV energy systems. Sol Energy 61(2):77–87
9. Ulleberg O (2004) The importance of control strategies in PV-hydrogen systems. Sol Energy 76:323–329
10. Vosen SR, Keller JO (1999) Hybrid energy storage systems for stand-alone electric power systems: optimization of system performance and cost through control strategies. Int J Hydrogen Energy 24:1139–1156
11. Manwell JF, McGowan JG (1993) Lead acid battery storage model for hybrid energy systems. Sol Energy 50(5):399–405

Chapter 5
Control of Hybrid Energy Systems

5.1 Hybrid Fuel Cell/Battery Controller Logic

There is growing interest in integrating fuel cells with batteries or supercapacitors to create stand-alone generators. These could be used in applications where diesel generators are commonly used today. The reason for wanting to do this is to reduce system cost. If one wanted to use a fuel cell only to meet a user demand that averages 250 W, for example, but that can reach short-term maxima of 1 kW, the fuel cell would need to be sized in order to meet that maximum demand, see Fig. 5.1. Capacity utilization of the fuel cell would be poor. By supplementing the fuel cell output with a battery bank, the fuel cell can be sized to meet the average user demand while the batteries meet the larger, short-term demands. As a result, significant cost savings compared with the fuel cell only system can be achieved. Additionally, by using the battery, the fuel cell can be buffered from transients in the load, thereby extending the fuel cell life [1].

This section will describe the logic that will be used to control a PEM fuel cell/battery hybrid system. The controller logic developed will be able to respond to three different load scenarios. The controller will also be tuned to buffer the fuel cell from load transients. The following block diagrams show the logic conceptually and the circuit switching that would be used in a digital system in order to achieve the desired system behavior [2].

In the first scenario, when there is no load and the battery is not fully charged, the controller will turn the fuel cell on and direct current to the battery. See Figs. 5.2 and 5.3.

In the second scenario, if the battery is not fully charged and the load is less than that of the rated output of the fuel cell, then the remainder of the fuel cell output will be directed to charge the battery. See Figs. 5.4, 5.5 and 5.6.

In the third scenario, if the load is higher than the maximum output of the fuel cell, the battery will supplement the fuel cell to meet the demand. See Figs. 5.6 and 5.7.

Lastly, if a sudden change in demand occurs, the system will disconnect the fuel cell and the battery will meet the demand for a short period. See Figs. 5.8 and 5.9.

S. Al-Hallaj and K. Kiszynski, *Hybrid Hydrogen Systems*,
Green Energy and Technology, DOI: 10.1007/978-1-84628-467-0_5,

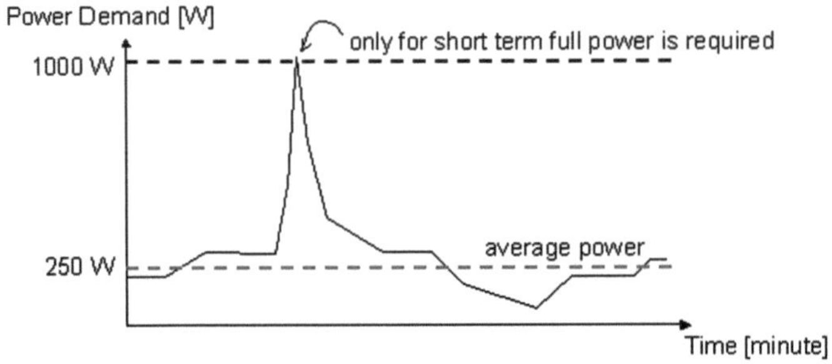

Fig. 5.1 Sample load profile illustrating short term, high power demand

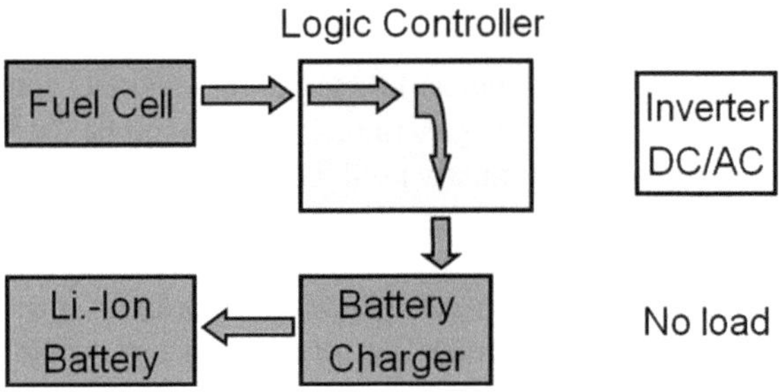

Fig. 5.2 Block diagram of system under no load with battery that is not fully charged

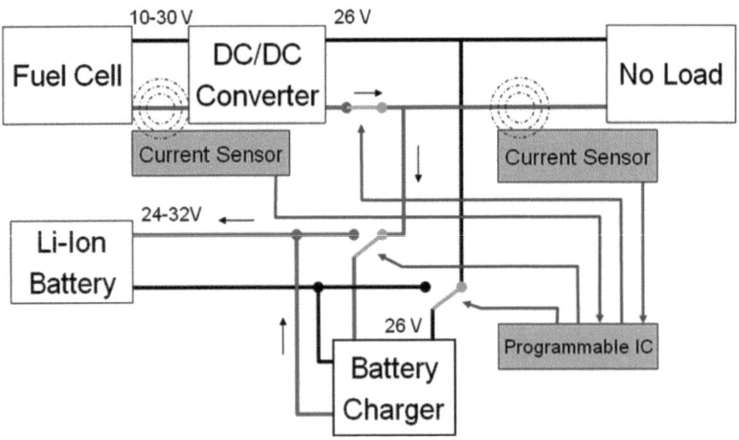

Fig. 5.3 Block diagram of system under no load with battery that is not fully charged

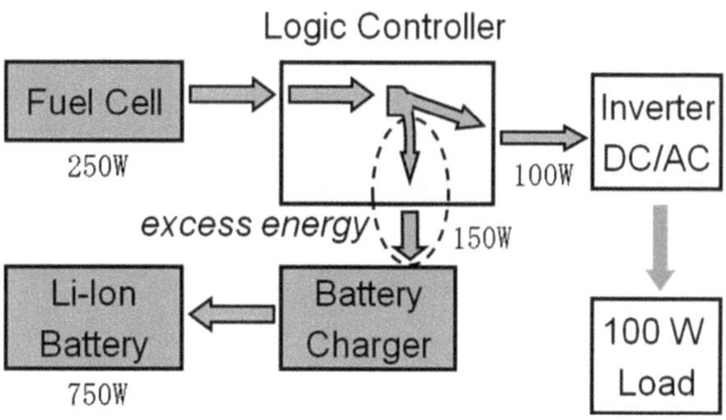

Fig. 5.4 Block diagram of system supply power to 100 W load using excess fuel cell capacity to charge battery

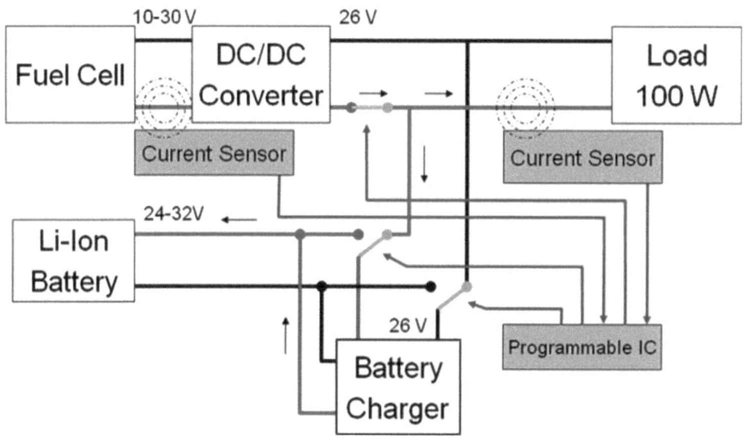

Fig. 5.5 Block diagram of system supply power to 100 W load using excess fuel cell capacity to charge battery

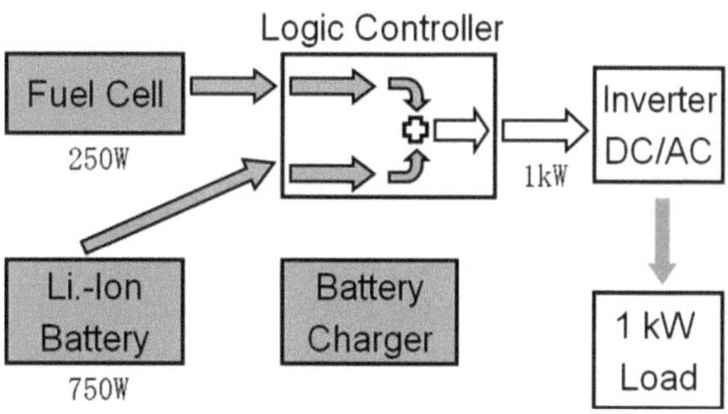

Fig. 5.6 Block diagram of system supply power to 1 kW load using both fuel cell and battery bank

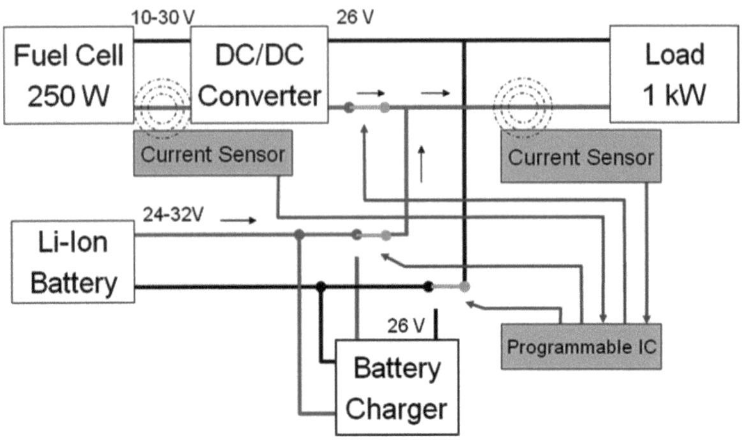

Fig. 5.7 Block diagram of system supply power to 1 kW load using both fuel cell and battery bank

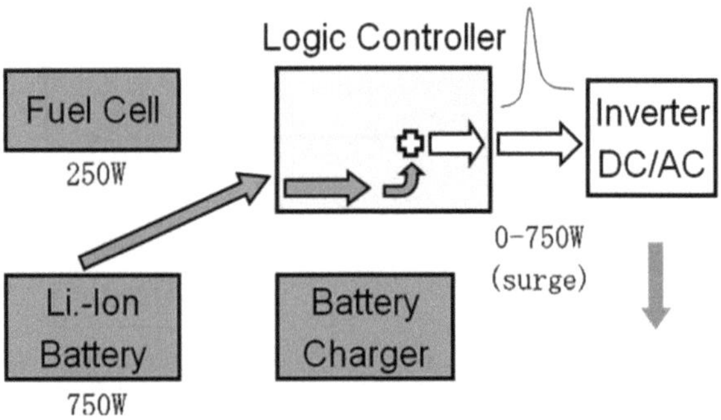

Fig. 5.8 Block diagram of system supply power to demand spike using battery bank only

5.2 Controller Design

The system can also be designed without the use of switches. The controller will be tuned to achieve the same behavior as described above. It is designed so that the battery will supply power to the load during the start up and in case of a sudden rise in the load demand. Thereafter, and while the fuel cell slowly reaches steady state, the battery slowly reduces the amount of power it supplies. The control diagram is shown in Fig. 5.10.

The feedback control strategy will work as described below [3]:

1. The demand load (P_L), the power output of the fuel cell (P_F) and battery power output (P_B) are measured.
2. The measured output from the fuel cell is subtracted from the limited load demand set-point, P_L^{SP}. The saturation block is included in the control loop to

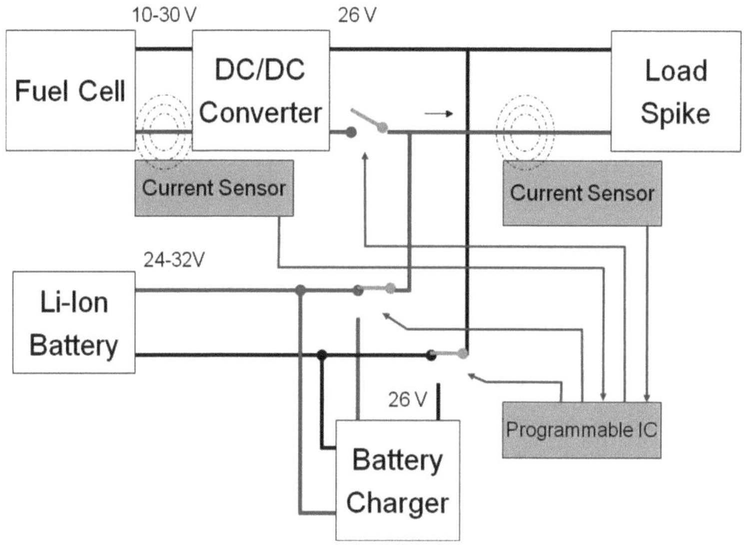

Fig. 5.9 Block diagram of system supply power to demand spike using battery bank only

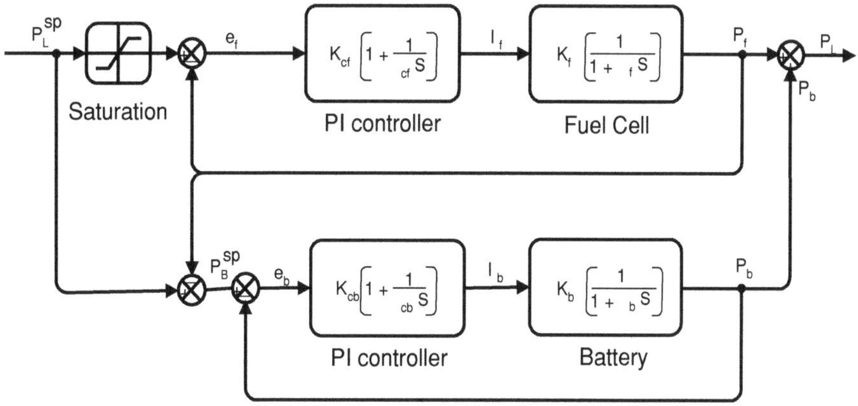

Fig. 5.10 Hybrid system control loop structure block diagram

limit the fuel cell output to its rated capacity. The fuel cell feedback error signal (e_f) is sent to the fuel cell PI controller.

3. The measured output from the fuel cell is also subtracted from the P_L^{SP}, which provides the battery set-point value, P_B^{SP}.
4. The battery feedback error signal (e_b) is determined by subtracting the measured battery output (P_B) from P_B^{SP}. The feedback error signal is then sent to the battery PI controller.

5. Outputs of the fuel cell and battery are combined and supplied to the load. However, during sharp change in the load demand, a time delay must be implemented to the fuel cell so that the battery responds faster to meet the load demand.
6. The demand load (P_L) is measured again and the procedure repeats itself.

5.2.1 DC/DC Converters

As shown in Fig. 5.10, a DC/DC converter is employed to condition the power output of the fuel cell. A bi-directional DC/DC converter is used to control the charging and discharging of the battery bank. Both DC/DC converters can be controlled in a manner such that sudden changes in the demand load will be absorbed by the battery, providing ample time for the fuel cell to respond. The bi-directional converter must supply the load using the battery when the fuel cell has not reached its steady-state condition. On the other hand, the converter must recharge the battery under normal conditions when the fuel cell feeds the load.

5.2.2 Open-Loop Modeling

5.2.2.1 Fuel Cell Characterization

Specifications of the fuel cell used in this example are shown in Table 5.1. The polarization curve of the fuel cell is shown in Fig. 5.11.

The fuel cell response to a step change from 2 to 10 amps is shown in Fig. 5.12. It exhibits a first order dynamic response. The first order dynamic response is described by Eq. 5.1:

$$\frac{di_f}{dt} = -\alpha_f i_f \tag{5.1}$$

where i = fuel cell current, α_f = exponential decay rate

Table 5.1 Fuel cell specifications

Manufacture	H-power
Type	PEM
Membrane	Polymer electrolyte
Reactants	Hydrogen and air
Open circuit voltage	36 V
Operating voltage	27.5 V at 11 amp
Operating current	11 A Nom. 14 A Max.
Rated power output at 11 A	302.5 W
Peak power	375 W at 14 A

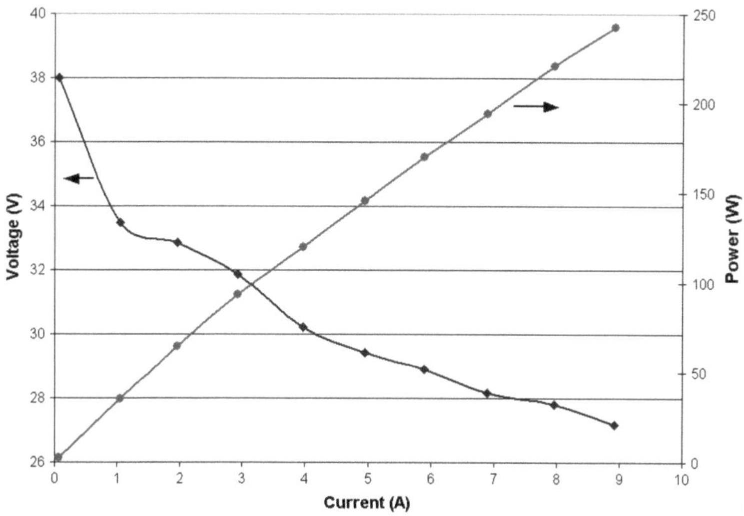

Fig. 5.11 Polarization curve of the fuel cell

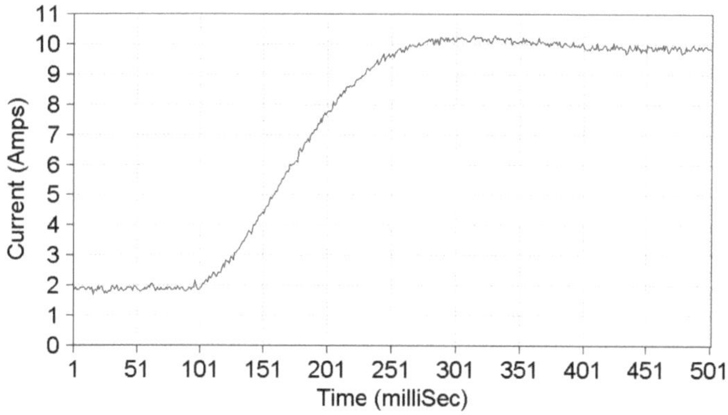

Fig. 5.12 Fuel cell response to step change in load current

Taking the Laplace transform, this gives Eq. 5.2:

$$I_f(s) = \frac{1}{\frac{1}{\alpha_f}s + 1} \tag{5.2}$$

Tests conducted on the fuel cell will determine its time constant, τ_f, which is simply the inverse of the exponential decay rate. The time constant informs the designer of how quickly the system responds to changes in input. The smaller the time constant, the more quickly the system will respond. An electronic load can be connected to the fuel cell stack to perform step changes in current demand.

The time constant is defined as the time the system takes to go from 0 to 63.2% of the final value. In this case, the final value is 7.06 A. This was calculated by adding the initial value to 63.2% of the magnitude of the step change. By analyzing the data, the time constant is determined to be 90 ms.

The gain, K_f, is chosen to match the power output characteristics. As a result, the first order dynamic model is described by Eq. 5.3:

$$\frac{P_f}{I_f^{sp}} = G_f(s) = \frac{K_f}{\tau_f S + 1} \tag{5.3}$$

$$K_f = 30\,V \quad \tau_f = 90\,mS$$

5.2.2.2 Battery Characterization

The same tests described above for the fuel cell can be repeated for testing of a single Ni–Cd battery was conducted to determine its time constant. The specifications of the battery bank used in this example are shown in Table 5.2. The battery is connected to an electronic load through a shunt of known resistance. The function of the shunt is to measure the current by measuring the voltage drop across it.

The battery data also suggests a first order response to the step change. The battery current response to a step change from 2 to 10 amp is shown in Fig. 5.13. The time constant is determined to be 100 ms.

Repeating the same procedure and fitting the data into the control loop structure, an open loop system with a first-order dynamic response was obtained. As a result, the first order dynamic model for the battery is given by Eq. 5.4:

$$\frac{P_b}{I_b^{sp}} = G_b(s) = \frac{K_b}{\tau_b S + 1} \tag{5.4}$$

$$K_b = 48\,V \quad \tau_b = 100\,mS$$

5.2.3 Tuning the Controller

This section will describe the tuning of the controller for the hybrid system. The results obtained from the above experiments are necessary to begin tuning the fuel

Table 5.2 Battery bank specifications

Manufacture	Alcad
Type	Ni–Cd
Composition	Cd in negative plate, Ni in positive plate
Nominal discharge voltage	1.2 V per cell
Total peak voltage	48 V
Total capacity	300 A-h

cell and battery PI controllers. The time constants for the fuel cell and battery as well as the measured value for the gain are known. Through a trial and error procedure, the tuning parameters are chosen. Open-loop and closed-loop battery response is shown in Fig. 5.14. The dotted line is the open loop response and the solid line represents the closed loop response. The final results of battery controller constants are shown below.

$$K_{cb} = 1 \quad \tau_{cb} = 0.9$$

The same procedure is repeated for the fuel cell. A plot of the fuel cell's open-loop and closed-loop responses is shown in Fig. 5.15. The dotted line shows the open loop response and the solid line shows the closed loop response. While experimental results display rapid response, it is desired that the fuel cell respond

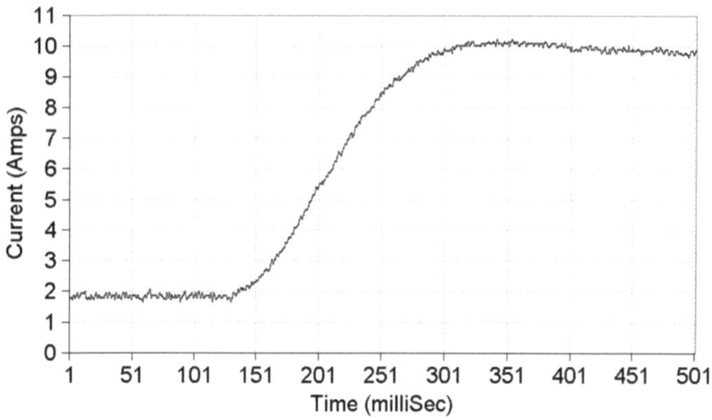

Fig. 5.13 Battery response to step change in load current

Fig. 5.14 Open-loop and closed-loop battery response

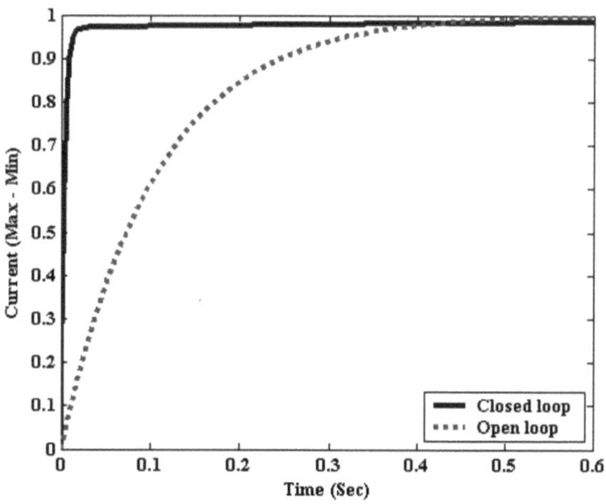

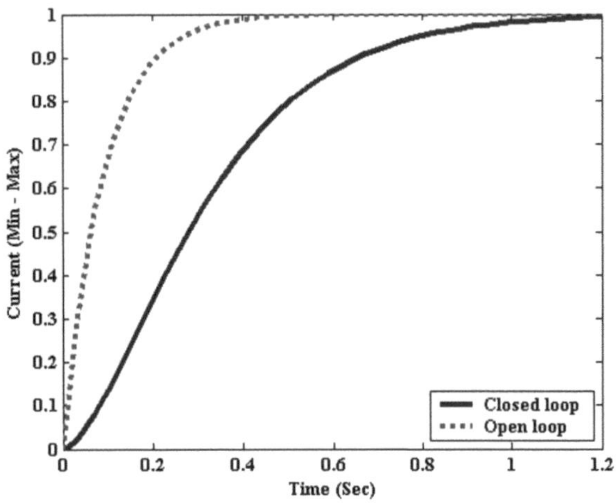

Fig. 5.15 Fuel cell response in open-loop and closed-loop configurations

slowly during a sudden step change so that the fuel cell life expectancy is prolonged. Therefore the fuel cell is allowed to take its time to reach steady state during quick start up. The plot is in agreement with the behavior desired for this fuel cell/battery control design. The controller fuel cell constants are adjusted and the final results are as shown below.

$$K_{cf} = 0.001 \quad \tau_{cf} = 0.01$$

5.2.4 Controller Logic with Battery State-of-Charge Considerations

State of charge (SOC) of the battery must be taken into consideration in order to prevent over discharge and ensure that the available battery capacity is well utilized. There have been many methods proposed for measuring the state of charge of batteries. One widely used method is the Coulometric method. This method calculates the ampere-hours either going into or coming out of the battery bank assuming a constant battery capacity. This method will be assumed in the design of the controller with SOC considerations.

In order to maintain the battery SOC within certain specified limits, it should be included in the control loop structure. The block diagram of complete control structure is shown in Fig. 5.16. Let's examine several cases to understand the control logic.

1. When the battery SOC is between SOC_{SPU} (upper set point) and SOC_{SPL} (lower set point) and more power than can be provided by the fuel cell is demanded, the battery will meet the remaining demand load.

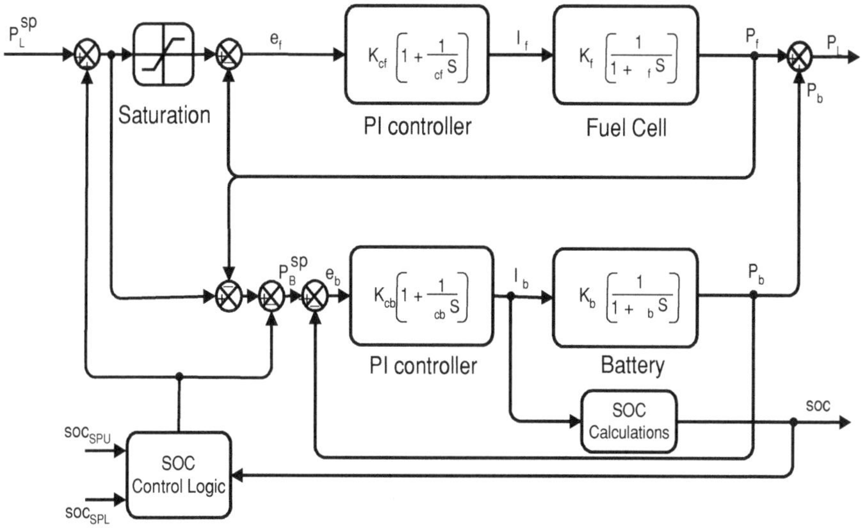

Fig. 5.16 Hybrid system control loop structure with battery SOC considerations

2. When the battery SOC is below SOC_{SPU} and the fuel cell is not running at full output, the controller will force the fuel cell to run at full output until the battery is fully charged.
3. When the battery SOC is above SOC_{SPU}, the controller will force the fuel cell to output less than is necessary to meet the demand load in order to discharge the battery until it reaches SOC_{SPU}.
4. When the battery SOC is below SOC_{SPL} and the fuel cell cannot meet the demand load, the system will fail to meet the demand load.

5.2.5 Simulation Results

The system response to a sudden increase in the load demand from 0 to 200 W is shown in Fig. 5.17. Initially the battery responds quickly. It then gradually drops supplying load as the fuel cell power output increases. The load output power is the summation of the fuel cell and battery output power.

When the demand load is higher than the fuel cell's maximum power output, the battery meets the rest of the demand load. As shown in Fig. 5.18, as the load power demand increases suddenly from 0 to 400 W, the battery supplies the demand load. As the fuel cell output increases, the battery output decreases. The fuel cell settles at its rated power output.

The system response to pulsating load changes is shown in Fig. 5.19. For the first 2 s, the system response is the same as explained earlier. For the next 2 s, when the load demand drops, the fuel cell reduces its power output. In the mean time,

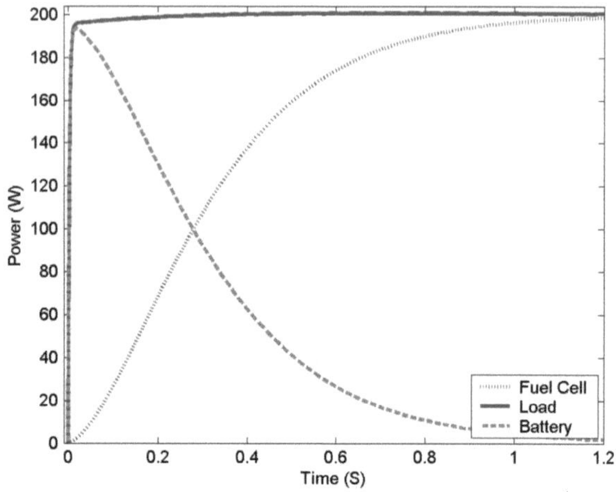

Fig. 5.17 Simulated system response for step change in load from 0 to 200 W

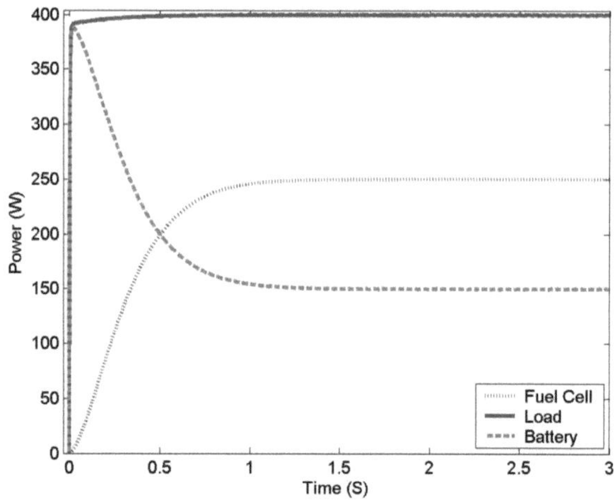

Fig. 5.18 System response to step change in load higher than fuel cell capacity

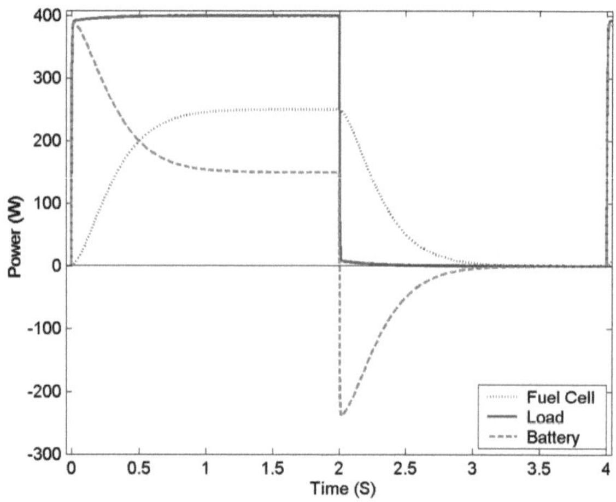

Fig. 5.19 System response to pulse load when the battery SOC is not considered

the excess power generated by the fuel cell will be used to charge the battery. This gives the fuel cell sufficient time to reach steady state. Battery discharging is considered positive and charging is considered negative on the *Y*-axis of the graph.

As shown in Fig. 5.20, if the battery SOC is less than the lower set point, the fuel cell generates more power than required by the load and this excess power is used to charge the battery.

The hybrid system response to a variable load profile is shown in Fig. 5.21. As the initial state of charge of the battery is below the lower set point, the controller operates the fuel cell at its maximum power level. The fuel cell produces more power than required by the demand load. This extra power generated will be used to charge the battery. A slight increase in the battery state of charge can be noticed over a 20-min period.

Fig. 5.20 System response to pulse load when the battery SOC is considered

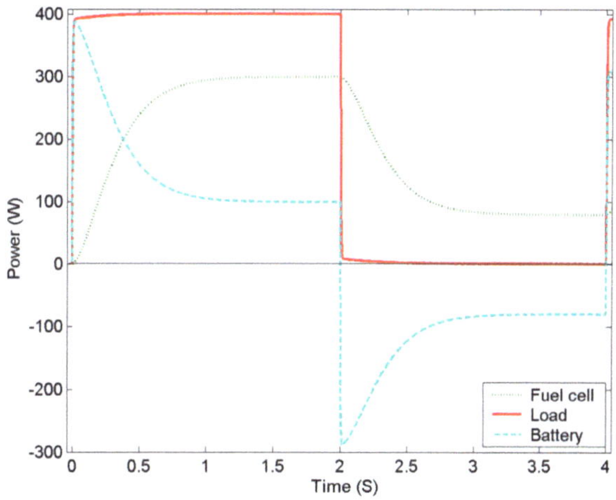

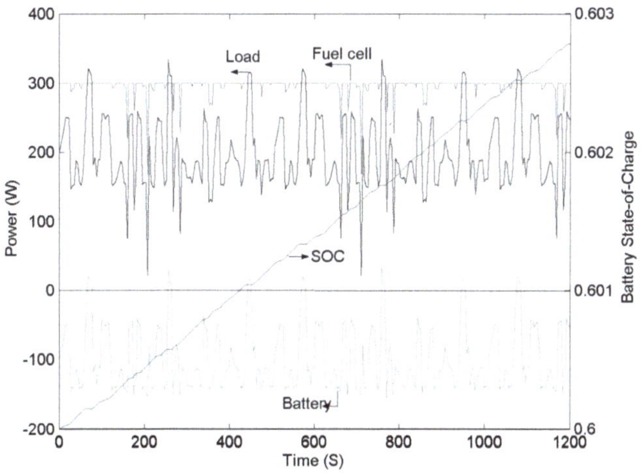

Fig. 5.21 Hybrid system response to a variable load profile

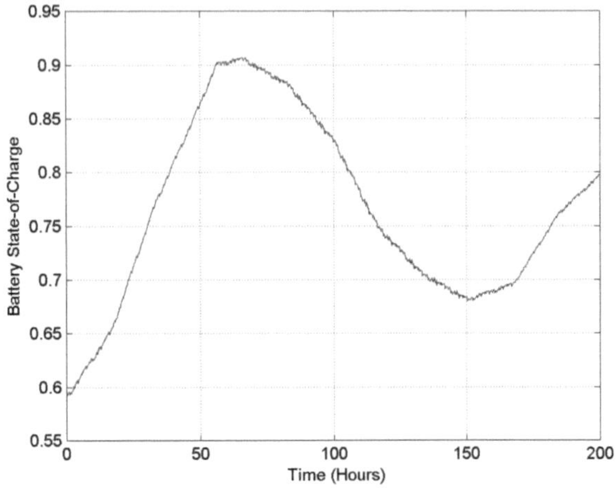

Fig. 5.22 Battery state of charge over 200-h system simulation

State of charge of the battery bank for a 200-h long system performance is shown in Fig. 5.22. The upper and lower SOC limits are set to 0.9 and 0.7, respectively. As the initial SOC is lower than 0.6, the fuel cell charges the battery. After the battery SOC reaches 0.9, it remains in the range between the upper and lower limits.

In this chapter, the control considerations of a fuel-cell battery hybrid system were presented. Protection of the fuel cell stack from large load changes and rapidly fluctuating loads is important for long fuel cell life. In addition, the reader was presented with the steps required to test component performance and to properly tune the controller to arrive at the desired system performance characteristics.

References

1. Uchimura M, Kocha SS (2007) "The Impact of cycle profile on PEMFC durability", session: Proton Exchange Membrane Fuel Cells (PEMFC 7). 212th electrochemical society meeting, Washington, DC, 7–12 Oct 2007
2. Yamagishi E, Al-Hallaj S (2006) Design and integration of hybrid fuel cell and lithium-ion battery system. Canadian Chemical Engineering Conference, Sherbrooke, Quebec, Canada, 15–18 Oct 2006
3. Nasiri A, Rimmalapudi VS, Emadi A, Chmielewski DJ, Al-Hallaj S (2004) Active control of a hybrid fuel cell-battery system. Power electronics and motion control conference, 2004. Conference proceedings, IPEMC 2004. The 4th international, 14–16 Aug 2004

Chapter 6
Case Study: Hybrid PEM Fuel Cell/Li-ion Battery System for a Non-Idling Airport Ground Support Vehicle

Chapters 1–3 introduced the major existing renewable energy technologies. Chapters 4 and 5 discussed how to combine these technologies into renewable hybrid energy systems. The case studies that follow in Chaps. 6 and 7 will illustrate how to apply this knowledge to real world problems.

6.1 Chapter Objectives

The case study presented in this chapter describes the design and implementation of a hybrid system for the elimination of engine idle in airport ground support vehicles. We will first examine the decision-making process by considering several combinations of renewable energy technologies. This will be followed by a discussion of why a PEM (Polymer Electrolyte Membrane) fuel cell/lithium–ion battery system is the most suitable design for this project. We will then discuss the details of the proposed design. The chapter will conclude in a discussion of the project results and a summary of the successes and limitations of the project along with proposed future work.

6.2 Problem Definition and Motivation

Airport ground support vehicles are used to escort and monitor contractors and visitors, primarily outdoors on the tarmac. Vehicle operators typically remain in the stationary vehicle during monitoring operations. The stationary vehicles, often trucks of the Ford F-150 size class, must therefore provide enough power to operate the air conditioning, emergency lighting, and communication radio systems. This is currently accomplished by idling the vehicle's engine (running the engine while the vehicle remains stationary). Engine idling results in increased fuel consumption, operating costs, and emissions of greenhouse gases and harmful

S. Al-Hallaj and K. Kiszynski, *Hybrid Hydrogen Systems*,
Green Energy and Technology, DOI: 10.1007/978-1-84628-467-0_6,

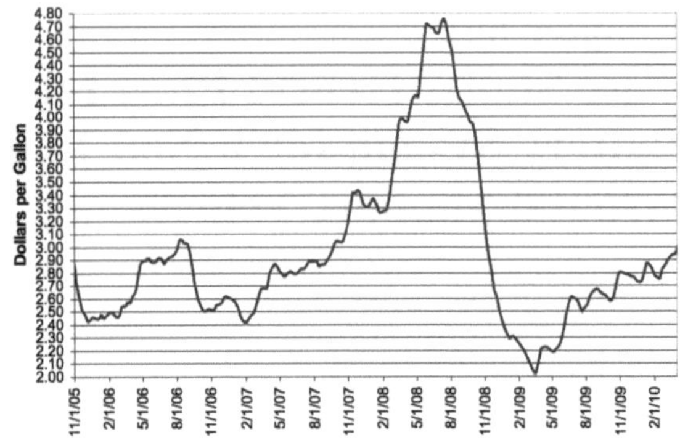

Fig. 6.1 Weekly US on-highway retail prices average all types [1]

pollutants. The project described in this case study was launched during a time when the price of US diesel was steadily increasing; between November 2005 and May 2008, the average price of diesel in the United States had risen from $2.7/gallon to $4.3/gallon (see Fig. 6.1) [1]. Shortly thereafter, the price of diesel steadily dropped from around $4.7 per gallon in mid-2008 to around $2.00 dollars per gallon in March 2009; however, as can be seen from Fig. 6.1, the price of diesel continues to rise after March 2009.

The undesired effects of engine idling can be reduced by designing a renewable hybrid energy system for the vehicle. The next section will outline the process of selecting an appropriate hybrid system.

6.3 Selection of Appropriate Hybrid System

For this project, it is desirable to use a fuel cell in conjunction with a backup power source (as opposed to using a stand alone fuel cell). Fuel cells are generally slow in response to load changes, and are expensive. For these reasons, a stand alone fuel cell system is not feasible. However, as shown in Chap. 5, using a backup power source (e.g. a battery) in parallel to supplement the fuel cell will both allow the system to respond more quickly to load changes and eliminate the need to oversize the fuel cell for maximum load capacity.

There are four options available for the backup power supply: (1) auxiliary diesel generators, (2) solar panels, (3) lead acid batteries, and (4) lithium–ion batteries.

Auxiliary diesel generators are the most practical and least expensive source of back-up power. They are also readily available off the shelf. However, diesel generators are still fossil fuel dependent and therefore release harmful pollutants and greenhouse gases to the atmosphere.

The use of solar panels is impractical because of the current high cost of PV cells. Also, since solar cell systems depend on the charging of batteries to supply power, the intermittence of solar energy would necessitate oversizing the hybrid system.

This leaves two options for consideration: lead acid and lithium–ion batteries. Lead acid batteries have been mainly used in the large battery industry (i.e. automotives, backup power systems, industrial equipment) while Li-ion batteries have been the dominant chemistry for portable electronics applications. Recently, however, lithium–ion batteries have started competing in markets traditionally controlled by lead acid batteries. More importantly, it has become clear that some industries need to transition to using Li-ion batteries if industry goals are to be met (e.g. the electric vehicle industry's goal of achieving a 200 mile range with a single charge).

Table 6.1 summarizes the comparative advantages and disadvantages of using lead acid or Li-ion batteries.

As can be seen from Table 6.1, there are several performance advantages to using Li-ion batteries: higher energy and power density, longer cycle life, greater efficiency, faster charge time, and longer shelf life.

These advantages significantly outweigh the disadvantages of using a Li-ion battery bank for backup power. Based on Table 6.1, the major disadvantages to using Li-ion batteries are: (1) the threat of thermal runaway, (2) higher battery costs, (3) insufficiently developed recycling methods, and (4) relative technological immaturity. First, the risks associated with thermal runaway can be reduced through the use of packaging materials. In this project, battery cells will be encapsulated with a phase change material (PCM)/graphite matrix for passive cooling and elimination of thermal runaway; this will be elaborated upon in Sect. 6.7. Second, for the purposes of this project, the higher cost of Li-ion batteries is justified by their considerably longer cycle lives. Third, the reason

Table 6.1 Comparison of lead acid and Li-ion batteries

	Lead Acid	Lithium–ion
Energy density	3–5 times lesser	3–5 times greater
Power density	5–60 times lesser	5–60 times greater
Cycle life	Typically less than 300 cycles	2,500 cycles (auto-grade) 700 cycles (laptops)
Capacity efficiency	As low as 50%	Never below 85% Often greater than 95%
Speed of charge	8 h	As fast as 80% in 6 min (charge speed can affect battery life)
Safety	Electrochemically safe Large mass (potential injuries)	Thermal runaway
Environmental impact	Solids: highly recyclable Liquid acid: toxic	Expected to be recyclable Expected aftermarket
System compatibility	Do not require electronics Less ready for data logging	Require integrated circuits Ready for data logging
Cost	$150/kWh	$1000/kWh
Shelf life	Typically 1 or 2 years	3–5 years Up to 10 years projected
Maturity	Longer commercial use	Young (since 1991)

recycling methods are not yet well developed for Li-ion batteries is that the technology is still relatively young. However, it is reasonable to expect that the majority of Li-ion battery components will be recyclable and that there will be an aftermarket for the reuse of Li-ion batteries. Finally, it is important to recognize that there remains much research to be conducted on Li-ion battery materials and manufacturing techniques before maximum performance can truly be achieved. However, for now, the performance benefits still outweigh using lead acid batteries.

As far as fuel cell type, it seems most logical to use a polymer electrolyte membrane (PEM) fuel cell. PEM fuel cells are considered the leading fuel cell type for passenger vehicle applications, largely due to their fast startup times and high power to weight ratios.

6.3.1 PEM Fuel Cell/Li-ion Battery Hybrid System

The hybrid system selected will consist of a polymer electrolyte membrane (PEM) fuel cell and a custom-designed, high-power Li-ion battery system. A bank of supercapacitors will be used to supply power during brief periods in which the fuel cell fails and the battery has not yet been activated. The system will also include a controller designed to route power throughout the system, and a control panel for monitoring battery voltage level, fuel cell status, and power consumption. A block diagram of the overall system configuration follows in Fig. 6.2 below.

The three sections that follow will outline the chosen design strategy and address how this strategy arises from the project goals and design specifications.

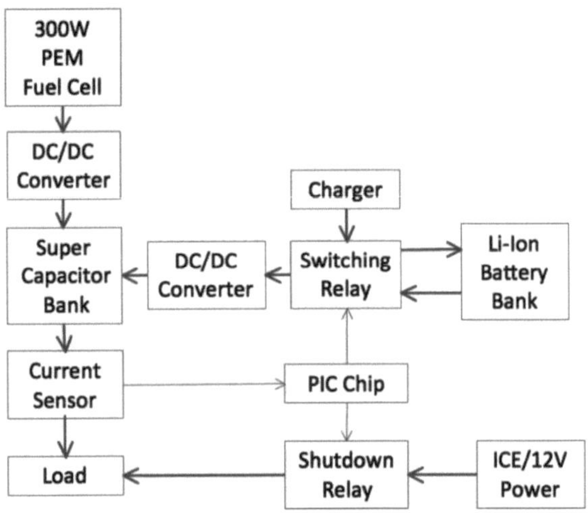

Fig. 6.2 Block diagram of PEM fuel cell/Li-ion battery system

6.4 Project Goals

The major goal of this project is to develop a renewable energy solution that is capable of powering an airport ground support vehicle's communication radio (3 A peak load), emergency lighting (7 A peak load), and air conditioning (80 A peak load) units without idling the vehicle's engine. The chosen solution must be able to provide power throughout the entire duration of a typical 8-h shift. The hybrid system should also be capable of automatically switching its power source to the vehicle battery in the event of failure. The selected PEM fuel cell/Li-ion battery system must accomplish all these goals.

6.5 Design Specifications

The PEM fuel cell/Li-ion battery system must also adhere to the following design specifications. The system must be able to:

1. charge battery cells on single-phase 120 V AC circuits,
2. charge battery cells on three-phase high voltage AC circuits,
3. utility charge batteries using excess fuel cell power, and
4. utility charge batteries while the internal combustion engine is running.

6.6 Design Strategy

This section outlines the specific design strategy chosen for the airport ground support vehicle's hybrid system. For continuous low-power loads (less than or equal to 250 W), the system will be entirely powered by the fuel cell. For intermittent and high-power loads (greater than to 250 W), the system will be supplemented with a high-energy Li-ion battery system which will supply 80% of power while the fuel cell continues to supply the remaining 20%.

An embedded, programmable interface controller (PIC) was also designed to monitor and switch battery states. This intelligent routing of system power will help to reduce the size and cost of the fuel cell, and to extend the life of the system. The controller will actively maintain the power output of the fuel cell at 250 W unless the load requires more power. When the power requirement exceeds 250 W, the controller will use the Li-ion battery to supply additional power, thereby keeping the fuel cell within necessary operating conditions.

The embedded PIC will also protect (cut off) batteries during under- or overvoltage conditions. The battery system will also be cut off during overcurrent conditions (when the power exceeds 1.5 kW).

Table 6.2 Li-ion battery module specifications

Cell type	High energy 18650
Cell configuration	7 Series, 5 parallel
Operating voltage	21–29.4 V
Nominal voltage	25.9 V
Nominal capacity	12 Ah
Maximum discharge current	12 A continuous
Peak discharge current	20 A for <1 s
Maximum charge voltage	29.4 V
Maximum charge current	6 A
Operating temperature range	−20 to 60°C
Charging temperature range	0–45°C

A control panel system was also be designed to provide separate real-time controls of the entire system and the fuel cell, and real-time monitoring of system and fuel cell on/off statuses, battery voltage level, and power consumption. An audible safety alarm was also incorporated into the control panel system.

6.7 Design Components

The details of each key component of the non-idle system are described throughout this section.

6.7.1 Li-ion Battery Pack Design

A Li-ion battery pack with 25.9 V/120 Ah ($\sim$3.2 kWh) was custom designed and built for this project. Within each battery pack ten 25.9 V/12 Ah Li-ion battery modules were used. In each module, 35 commercial cells Type 18650 (3.7 V/2.4 Ah) were configured as seven in series and five in parallel (7S $\times$ 5P). The cells were encapsulated with a phase change material (PCM)/graphite matrix for passive cooling. The PCM/graphite matrix absorbs and conducts heat away from battery

Table 6.3 Li-ion battery pack specifications

Modular battery configuration	1 Series, 10 parallel
Operating voltage	21.0–29.4 V
Nominal voltage	25.9 V
Nominal capacity	120 Ah
Maximum discharge current	120 A continuous
Peak discharge current	200 A for <1 s
Maximum charge current	60 A
Operating temperature range	−20 to 60°C
Charging temperature range	0–45°C

Table 6.4 Fuel cell specifications

Fuel cell type	PEM (air breathing)
Manufacturer	H-Power
Model	PS250
Fuel supply	Ultra-high purity hydrogen
Hydrogen pressure	30–50 psi
Hydrogen storage	Dual 900 L metal hydride tanks
Power output	250 W at 28 V
Operating voltage	~20–36 V

cells eliminating the need for an active cooling system. Other benefits of using the PCM technology include extending the lifetime and running time of batteries and eliminating thermal runaway. Specifications for the battery pack and individual battery modules used follow in Tables 6.2 and 6.3.

6.7.2 Fuel Cell

The PEM fuel cell will use ultra-high purity hydrogen stored in metal hydride tanks. Specifications for the fuel cell follow in Table 6.4.

6.7.3 Hydrogen Storage Tanks

Specifications for the metal hydride storage tanks follow in Table 6.5. In addition, Fig. 6.3 shows the relationship between power and run time for fuel cell stacks utilizing this metal hydride storage tank (Ovonics Model 85G250B) [2, 3].

6.7.4 DC/DC Converters

The non-idle system will use DC/DC converters to step down the voltages of both the fuel cell and Li-battery to the power source switching circuit. Specifications for the DC/DC converters follow in Table 6.6.

Table 6.5 Hydrogen storage tank specifications

Storage type	Metal hydride
Manufacturer	Ovonics
Model	85G250B
Diameter	3.5 in.
Length	15.1 in.
Weight	14 lbs.
Nominal capacity	900 L
Nominal discharge rate	600 W

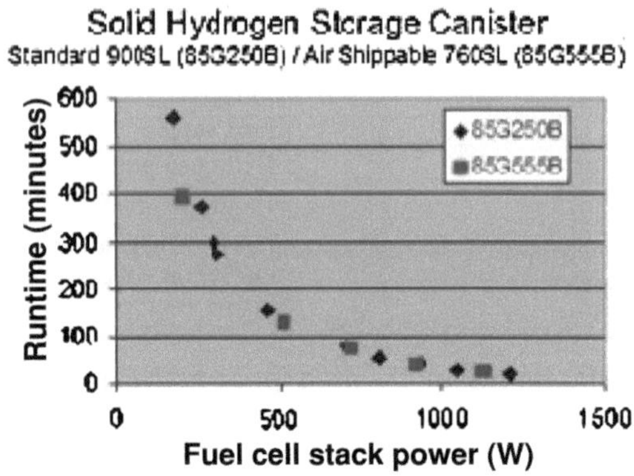

Fig. 6.3 Run time versus stack power for fuel cell storage tank model 85G2-250B

Table 6.6 Fuel cell DC/DC converter specifications

Usage	Fuel cell step down
Manufacturer	Newmar power
Model	48-12-35I
Input range	20–56 V
Output	13.6 V (variable)
Max current	20 A

6.7.5 Battery Charger

Specifications for the Li-ion battery charger follow in Table 6.7.

6.7.6 Control System: Control Panel and Controller

The control panel interface for the control system is shown below in Fig. 6.4. Note the four status updates indicated: (1) non-idle system status, (2) fuel cell status, (3) battery voltage level/utilization, and (4) power consumption.

When the non-idle system is enabled, the power source for the communication radio and emergency lighting systems will be routed to the hybrid system. The power

Table 6.7 Li-ion battery charger specifications

Battery type	Li-ion
Manufacturer	Delta-Q
Model	QuiQ 913-24xx
Nominal DC output	24 V
Max DC output	34 V
DC output current	25 A
AC input	85–265 V AC
AC input current	12 A @ 120 V AC

Fig. 6.4 Control panel interface

will be automatically rerouted to the vehicle battery in case of power loss. Note, however, that the AC unit used will be an auxiliary unit operating only from hybrid system power.

The Control Panel Interface (CPI) communicates with the control board in the energy compartment. The control board calculates real time power consumption and estimates battery state of charge (SOC). The controller also sends fuel cell status to the CPI. The control board has a very simple physical layout. A current sensor sends data to a Programmable Interface Controller (PIC) chip. This PIC chip decides when to switch the batteries from charge to discharge (via the relay board), and a bank of supercapacitors keeps the load powered during the short power switch. The controller can be in any of the states listed in Table 6.8 on the following page.

Figure 6.5 illustrates the details of the controller design.

6.8 Results

This section presents the results of the non-idling airport ground support vehicle project upon completion.

6.8.1 Data Acquisition Methods

The data that follows was acquired using an 8-channel analog voltage input module in conjunction with National Instruments Compact FieldPoint™, a programmable automation controller. The following data was recorded to allow for calculation of power consumption and an estimation of run time: (1) current sensor voltage, (2) fuel cell voltage, (3) Li-ion battery voltage, and (5) vehicle battery voltage.

Table 6.8 Possible controller states

State	Component	Description
Idle	Current sensor	Idle
	Super capacitor bank	Fully charged
	Relay board	Switched to charger
	Available power sources	Fuel cell, capacitor bank, charger power
0–300 W	Current sensor	0–300 W
	Super capacitor bank	Stabilizing fuel cell ripples
	Relay board	Switched to charger
	Available power sources	Fuel cell, capacitor bank, charger power
0–300 W (no fuel cell)	Current sensor	0–300 W
	Super capacitor bank	Fully charged (while battery has capacity)
	Relay board	Switched to discharge
	Available power sources	Battery bank, capacitor bank
300–1000 W	Current sensor	300–1000 W
	Super capacitor bank	Fully charged (while battery has capacity)
	Relay board	Switched to discharge
	Available power sources	Fuel cell, battery bank, capacitor bank
300–1,000 W (no fuel cell)	Current sensor	300–1,000 W
	Super capacitor bank	Fully charged (while battery has capacity)
	Relay board	Switched to discharge
	Available power sources:	Battery bank, capacitor bank
Switching state	Current sensor	Varies
	Super capacitor bank	Being discharged
	Relay board	N/A
	Available power sources	Capacitor bank, fuel cell
Off State	Current sensor	N/A
	Super capacitor bank	Slowly draining
	Relay board	Switched to charger
	Available power sources	ICE/12 V battery

6.8.2 Results and Discussion

The data collected and calculated is summarized in Fig. 6.6, which shows the behavior of the PEM fuel cell/Li-ion battery hybrid system under maximum load for a single run cycle.

Several observations can be made from the graph in Fig. 6.6. As expected, when the power load is less than or equal to 250 W, the fuel cell provides sufficient power and the Li-ion battery remains in the idle charging state. However, when the fuel cell loses power, the battery immediately becomes active. The graph also demonstrates that the fuel cell behaves sporadically when there is little to no power load, and gradually decreases its voltage during periods of high load. The fuel cell

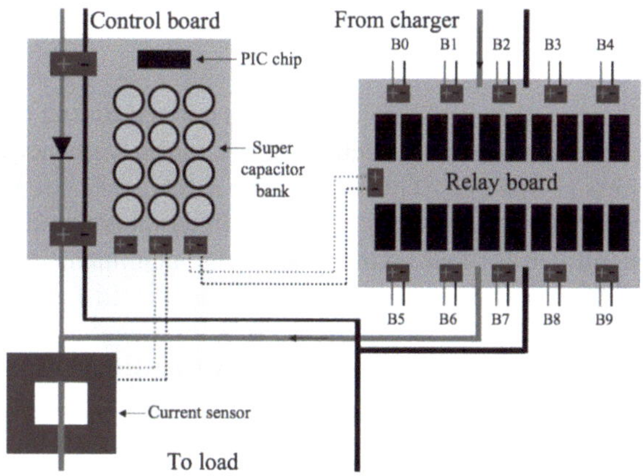

Fig. 6.5 Diagram of controller design

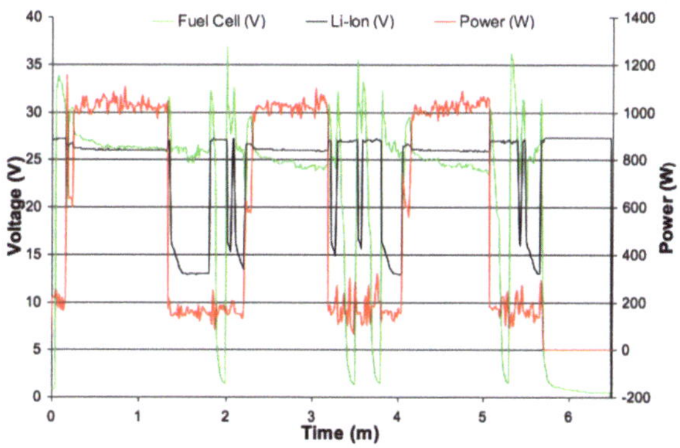

Fig. 6.6 Maximum load run cycle for fuel cell/Li-ion battery hybrid system

voltage peaked at 36.9 V while the power load peaked at 1175 W. Note that the periodic spikes in power indicate activation of the auxiliary air conditioning unit. Also, the graph seems to indicate that the supercapacitors adequately handled the current draw in the short period of time during which the fuel cell failed and Li-ion battery had not yet been activated.

6.8.3 Calculations

The following calculations were made from the data. Since the Li-ion battery loses an estimated 1.08 V/h, the estimated run time for the battery and fuel cell system was calculated to be 8.4 h, assuming a voltage range of 28.7–19.6 V (or 4.1–2.8 V per cell). Additionally assuming a 20% loss of fuel cell power, the estimated run time for the Li-ion battery by itself was calculated to be 6.7 h. Thus, at 25 A, the Li-ion battery charger was calculated to fully charge the battery in 4.8 h.

6.9 Limitations and Pitfalls

There were several limitations encountered during the project. First, cost effective methods of refilling metal hydride tanks were not readily available. The compressed hydrogen tanks also pose safety risks for vehicle operators, and airport contractors and visitors. Another limitation is that the tanks need to be continually monitored for leaks, corrosion, and contaminants to the hydrogen supply.

Furthermore, the goal of charging the Li-ion batteries with excess fuel cell power and while the internal combustion engine is running were not yet achieved upon completion of the project. Also, the 1 h quick charge of batteries was not yet implemented.

Finally, an auxiliary A/C unit was installed on the vehicle roof to avoid the use of onboard air conditioning, which is operated by the engine. This task turned out to be quite a technical hurdle. It became necessary to oversize the air conditioning units due to the lack of available DC-only units. The A/C unit implemented during the project drew significantly more power than necessary, wasted resources, and even required major mechanical support as it threatened to collapse in the roof of the vehicle.

6.10 Plans for Future Work

The following suggestions for future work were made upon completion of the project: (1) a power consumption study to determine the hybrid power system run time and how utility charging and/or additional components can increase run time, (2) a fuel saving study comparing idling and non-idling airport ground support vehicles, (3) a feasibility study determining the pros and cons of continuing to use metal hydride storage tanks, (4) an investigation of the possibility of using to 5000 psi compressed hydrogen tanks to allow for quick recharge and longer run times, (5) a search for a more cost-effective, energy-saving A/C unit for future airport fleet vehicles, (6) an inquiry into higher voltage battery chargers for faster charge time, and (7) enhancing utility charging control by offloading charge control from the charger to the PIC controller; that is, allowing the PIC controller to redirect surplus power from the vehicle engine to the battery bank.

6.11 Conclusion

This project successfully accomplished its main goal: the development of a power system that provides adequate power to operate communication radio, emergency lighting, and auxiliary air conditioning systems in an airport ground support vehicle. Under maximum load, the PEM fuel cell/Li-ion battery hybrid system used was determined to last approximately 8.4 h when using the fuel cell and battery, and 6.7 h when using only the battery. The system developed was able to charge Li-ion batters in under 5 h on a normal 120 V AC, 15 A circuit. Also, as planned, the lighting and communication radio system automatically switch to the vehicle battery as a power source upon failure of the hybrid system.

References

1. "Weekly Retail On-Highway Diesel Prices." U.S. Energy Information Administration. http://tonto.eia.doe.gov/oog/info/wohdp/diesel.asp?featureclicked=2&. Accessed May 2008
2. http://www.fuelcellstore.com. Accessed May 2008
3. "FCT Fuel Cells: Types of Fuel Cells." U.S. Department of Energy: Energy Efficiency & Renewable Energy. http://www1.eere.energy.gov/hydrogenandfuelcells/fuelcells/fc_types.html. Accessed March 2010

Chapter 7
Case Study: A Hybrid Fuel Cell/Desalination System for Caye Caulker

7.1 Chapter Objectives

The case study presented in this chapter discusses the key decisions that factor into the design of a hybrid fuel cell/desalination (HFCD) system to supply a developing region with adequate electrical power and water. The focus of this case study is Caye Caulker, a Carribean island located off the coast of Belize. Caye Caulker has limited fresh water sources and currently uses diesel generators as its sole source of power. The goal of this case study is to replace these diesel generators with an HCFD system that can also provide Caye Caulker with potable water.

First, the selection of an appropriate hybrid system for Caye Caulker will be examined, taking the following into consideration: the variety of possible system integration configurations and optimization strategies, the advantages and disadvantages of different fuel cell and desalination technologies, and the results of simulation and economic analyses. This chapter will also present the design equations used in the simulation, and the theory that informs the economic analysis. The chapter will conclude with a detailed discussion of the proposed solution for Caye Caulker and its economic feasibility.

7.2 Motivation

As the world's population continues to increase rapidly, it becomes evermore important to make efficient use of finite resources. Two major problems currently facing fast-growing populations are lack of (1) an adequate fresh water supply and (2) sufficient electrical power. For example, in the Middle East/North Africa (MENA) region, water availibility often falls in the 0.3–0.6 m^3/day per capita range; the international standard for water scarcity is 0.5 m^3/day per capita [1].

S. Al-Hallaj and K. Kiszynski, *Hybrid Hydrogen Systems*, Green Energy and Technology, DOI: 10.1007/978-1-84628-467-0_7,

In addition, the total regional power demand (0.3 kW per capita in 2007) is increasing by 6% annually, versus a worldwide average of 3%.

Fortunately, early analysis indicates that hybrid fuel cell/desalination systems can be used to remedy these issues: A 100,000 m^3/day desalination plant can service between 170,000 and 330,000 people daily, provided the fuel cell system generates between 54 and 106 MW of power, in addition to the amount required for desalination. This solution is ideal for developing regions with moderate populations that are arid, near saltwater sources, and remote from an electrical grid. This includes many islands which currently rely on diesel generators for their power needs and small amounts of rainwater to for a potable water supply. The next section will outline the specific issues facing Caye Caulker, a small island off the east coast of Belize.

7.3 Problem Definition: Water Shortage in Caye Caulker

Caye Caulker receives 50" of rainfall annually, its sole source of potable water for a population of 1,000 residents and visitors. The island is powered by three Caterpillar diesel generators that have a combined capacity of 575 kW. The price of electricity on the island is $0.20/kWh, which is marginally greater than the cost of diesel fuel. For a typical diesel generator system, 17 kWh can be generated from one $3 gallon of diesel. The price of electricity is $0.17/kWh. The goal of this case study will be to replace the diesel generators with an appropriate hybrid fuel cell/desalination system.

7.4 Tools for Selecting Appropriate Hybrid System

7.4.1 Hybrid Fuel Cell/Desalination (HFCD) Systems: An Attractive Solution

Fuel cells were originally developed to supply electricity for specific applications. However, they also release a considerable amount of waste energy during operation in the form of hot water or steam. This waste energy can be captured by cogeneration systems to improve overall hybrid system efficiency. Utilizing this waste heat also helps reduce the cost of fuel cells. Additionally, the use of excess electricity generated by fuel cells for water desalination can eliminate the need of electricity storage in batteries during off-peak hours. The storage of water is easier, cheaper and more efficient than storing electricity, and, unlike electricity, the price of water does not fluctuate much. All these factors make fuel cells an attractive solution for stationary power applications where high quality and reliability are of utmost importance. Figure 7.1 shows a comparison of efficiency and capacity among various energy conversion technologies. Fuel cells not only have the

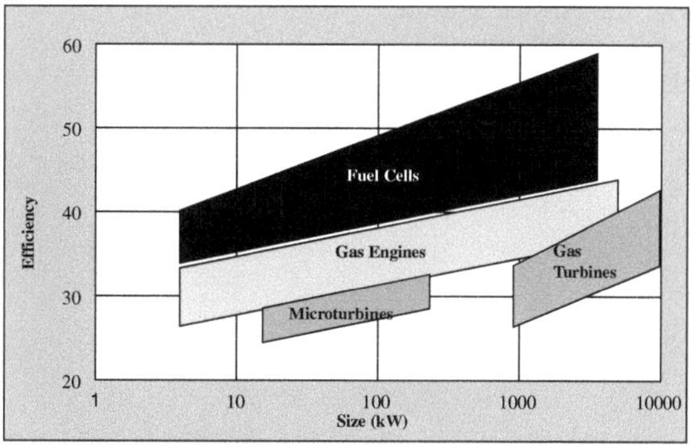

Fig. 7.1 Energy conversion efficiencies for various technologies [2]

highest efficiency but also the broadest range of capacity of the considered technologies.

To reap maximum benefit from the positive attributes of fuel cells, the HFCD systems should be used in situations where conventional desalination/cogeneration units have been shown to fail. As can be seen from reviews of current desalination/cogeneration plants, the industry standard is to create massive facilities. This makes sense, as the types of technologies used in these large facilities improve in performance and efficiency with size. These plants are also very sensitive to economies of scale. Fuel cell systems, on the other hand, are modular, which means that their efficiencies are not only independent of size, but also less sensitive to economies of scale. The same can be said of certain modular desalination methods, such as reverse osmosis (RO).

The modular nature of fuel cells and reverse osmosis also minimizes on-site construction. The fuel cells and RO units can all be built at another location, then reassembled wherever desired.

One issue with implementing a high technology product in an underdeveloped region is the lack of qualified operators. Currently, many fuel cells types require a high level of maintenance due to lack of reliability and robustness. As some of the research and development shifts focus from improving ideal performance to creating a viable product, the operation and maintenance factors should become less of an issue. Because fuel cells do not have any moving parts, the operation and maintenance has the potential to be very simple.

7.4.2 System Integration Configurations

When deciding how to hybridize a system, the ultimate goal is to provide the demanded outputs while maximizing profits. In a public utility project, this does not necessarily mean having a high rate of return but rather reducing the amount of money the public must pay to support the project. In many countries, water and electricity are subsidized, so a successful project results in decreased subsidies.

Even though there are only two elements defined in this system (a fuel cell and a desalination unit), there are several possible hybridizations, or integration

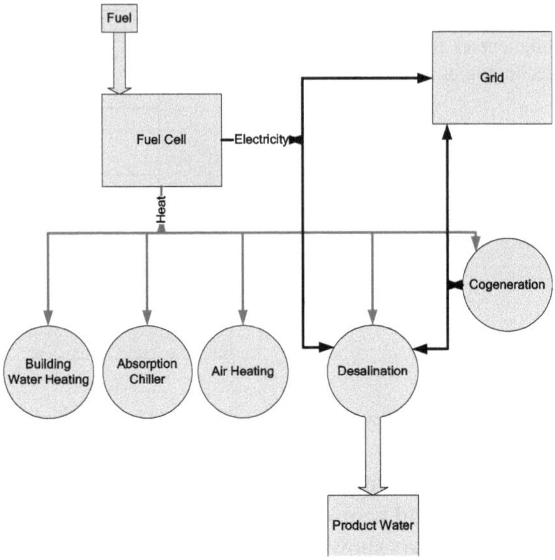

Fig. 7.2 Possible uses for exiting heat/electricity streams [3]

configurations. The fuel cell and desalination units are sized to meet the maximum production capacities for each respective product, but there will be times when the electric power demand will be low, and the fuel cell will not produce enough thermal waste to adequately power the desalination process. When this is the case, either (1) the fuel cell will have to overproduce in order to make more waste heat for desalination, or (2) some other method of heating the water must be used (e.g. a burner or electric heater).

When the water demand is low, the decision must be made to either (1) produce excess water that will be stored, or (2) devote the waste heat to power cogeneration units. Other possible uses of the excess heat include district water heating, air heating, or absorption chillers. The potential uses of system outputs (heat and electricity) are diagrammed in Fig. 7.2.

Depending on location and operating conditions, each system will have a different combination of components that optimizes performance. For example, in some situations, the water and power demands may be met while generating only a small amount of excess heat. In this case, the expense of cogeneration hardware might far outweigh the profit that could be derived from turning this small amount of excess heat into power.

The following three sections highlight key factors that can influence system integration configuration: (1) fluctuation in power and waste demand, (2) distance from electricity grid, and (3) the availability of fuel and water sources.

7.4.2.1 Fluctuation in Power and Waste Demand

As can be seen in the Fig. 7.3 showing monthly electricity demand for Abu Dhabi, there can be significant increases in power demand during summer months due to

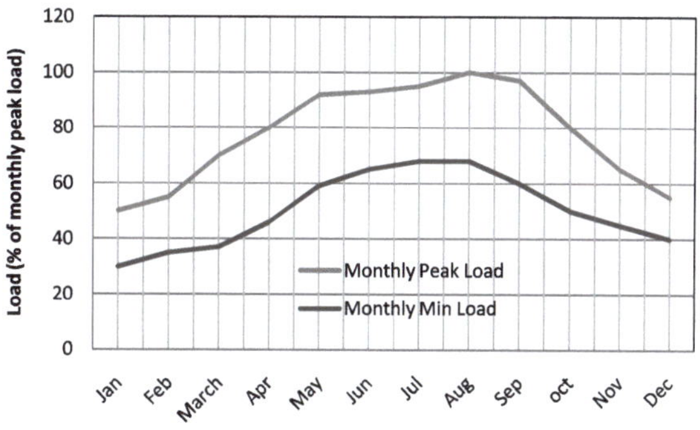

Fig. 7.3 Monthly electricity demand for Abu Dhabi [4]

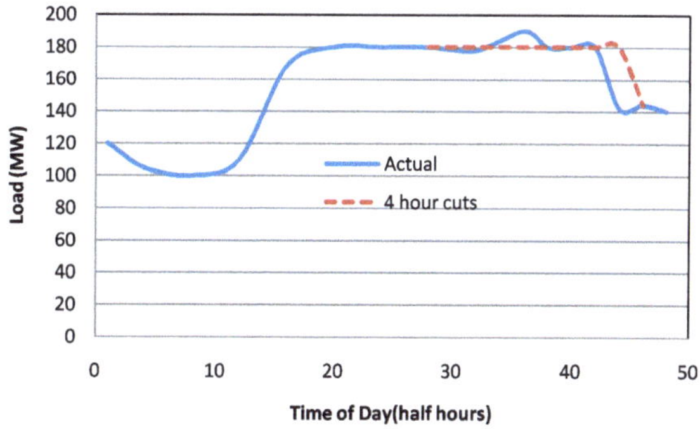

Fig. 7.4 Daily fluctuations in electric demand in New Zealand

higher temperatures. Water demand also exhibits a peak in summer months but it does not fall to the same extent in winter. Both seasonal and daily variations in power and water demand must be taken into consideration during HFCD design.

Also, unless the fuel cell system will supply only a small percentage of the local electric demand, it will have to be designed in accordance with grid load fluctuations. Throughout the day, the power demand for a region will often vary by 25–40% of the average load. Figure 7.4, showing electrical demand fluctuations in New Zealand, is representative of a typical grid load profile [5]. The average load is around 150 MW, with a maximum and minimum of 190 and 100 MW respectively.

Note that the power system represented in the Fig. 7.4 only averages 80% of its peak potential throughout the day. This concept must be applied to the relationship between desalination capacity and fuel cell plant size. If a desalination unit requires a fuel cell system providing 40 MW at full capacity for 24 h a day, the practical solution is to have a 50 MW system that will average 80% load throughout the day.

The red line is the average level at which the desalination unit must produce to meet the minimum daily requirements. The blue line represents the production if the desalination system is only sized to produce that average amount of water throughout the entire day. The green line represents the necessary desalination production in order to produce the same amount of water as the red line average. This graph means that the desalination plant must be sized to be able to produce the necessary amount of water if it is operating at approximately 80% capacity.

Due to the slow response time of desalination units, the production of water is not a function of instantaneous demand, but rather consistent throughout the day with some variation. The water will then be sent to a storage tank, so that there will always be enough on hand to meet any urgent demand. In any hybrid system that has two independent output demand curves, there will be times when storage is required. In this system, if 100,000 m^3 are to be produced each day, then either (1) 4,200 m^3 must be produced each hour, or (2) hours of higher production must be alternated with hours of lower production. Considering the fact that the fuel cell power demand will also be fluctuating with time, it makes sense to have the water production vary with it. This will reduce the amount of excess power being produced at any given time. The main problem with this strategy is that the desalination system must now be made larger than optimally necessary.

7.4.2.2 Distance from Electricity Grid

Another factor influencing integration decisions is the proximity of the HFCD system to the electricity grid. That is, whether the system is close to the grid, or far enough that it has to provide electricity and water independently from the grid. The HFCD system should be designed to cover peak demands for both water and electricity. During off-peak time, water can be easily stored for peak time use. However, storing electricity is much more difficult. If the HFCD system is located near the electricity grid, the electricity produced can be conveniently dumped into the grid.

The distance from the grid affects not only the choice of integration but also the mode of operation. For example, if electricity can be fed to the grid, the maximum available energy from fuel cells will be used to produce electricity during the peak period.

7.4.2.3 Availibility of Fuel and Water Sources

Another factor to be considered is the availability of a fuel source for the fuel cell. The choice of fuel cell is affected by the availability of natural gas or high-purity hydrogen. Fortunately, since the system relies upon a source of seawater, most of the target regions will have access to fuel via shipping, if fuel is not readily available through a pipeline. Alternatively, solar energy could also be used to generate hydrogen for fuel cells.

7.4.2.4 System Integration Methodology

The methodology that follows in Fig. 7.5 has been developed to help guide you through the multi-step process of choosing a suitable integration configuration for an HFCD system.

HFCD systems can be designed for a variety of applications such as providing electricity, water, and/or air conditioning to a building, military base, or a small village. However, the first step in selecting the appropriate HFCD system among the possible integrations is always to determine the water and power demand. Water and electricity demands vary from country to country, or even within the same country. For HFCD system design, the demand has to be estimated based on surveys that take future needs into consideration. The demand for electricity should be given greater significance than the demand for water, as variations in water demand are never as sharp as those in electricity demand and water storage is easier and cheaper than electricity storage. Hence, HFCD systems must be designed based on the peak electricity demand.

The next step in the methodology is to weigh possible fuel cell types and desalination processes against one another, taking into account the availability of potential water and fuel resources. One can then begin to look at possible integration options. Note that the simultaneous electrical and thermal energy needs for desalination will limit the possible fuel cell types for integration. The basis for differentiating between desalination processes will frequently be process efficiency, but this must also be weighed against economic considerations.

It is now possible to model several HFCD systems in order to determine the best modes of operation. In these models, power from fuel cells will be distributed

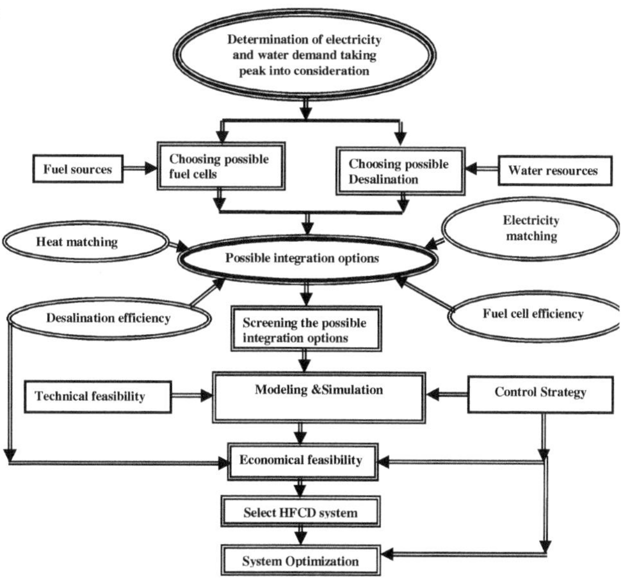

Fig. 7.5 Methodology for integrating fuel cells with desalination [6]

among the load and desalination unit. Power will also be distributed among any extra units in the process (e.g. air conditioning). These models give a better picture of the technical feasibility of potential HFCD systems. The simulations also help to develop control strategies for the HFCD systems (see Sect. 7.4.3).

However, one must always keep economic feasibility in mind as the most important criterion for selecting both fuel cell type and desalination process. The data from the simulation models should inform a thorough study of economic feasibility that takes into account both capital and operating costs. Only after this feasibility study is completed is it possible to select the HFCD system.

The final step is to optimize this selected HFCD system for maximum efficiency and economic benefit.

7.4.3 Control Strategy

Fuel cells have a tendency to work more efficiently when subjected to constant loads. The demand for electricity and water will not be constant throughout the day or during the year. Hence, it is convenient to design a robust system that will alternate between producing water and supplying electricity, while always keeping the fuel cell at optimum conditions. As mentioned previously, the storage of water is easier and cheaper than the storage of electricity. Therefore, it would be convenient to design the HFCD system based on the variation of the electricity demand, taking water demand into consideration.

The control strategy will determine the optimum utilization of the available fuel cell energy. Electrical energy output from the fuel cell is mainly distributed among the load and the water desalination unit, while thermal energy is mainly used for desalination. The design of the HFCD system can be based on the daily or seasonal variation in the electricity demand, as discussed previously. Of course, the control strategy of the HFCD system depends on the used fuel cell and desalination process, which therefore, depends on the application, system size, and the other factors discussed in previous chapters. Finally, the HFCD system must be flexible to switch between water or electricity production.

7.4.4 Comparison of Possible HFCD Technologies Using Methodology

This section will outline some key observations that arise from applying the methodology to the Caye Caulker case study. The data presented in this section will be used to help determine the best HFCD solution for Caye Caulker. First, however, a brief summary of various desalination technologies will be presented, along with a comparison of their relative advantages and disadvantages.

7.4.4.1 Summary of Desalination Technologies

Conventional desalination processes fall under two categories: (1) membrane desalination and (2) thermal desalination. The main membrane desalination process discussed in this section is Reverse Osmosis (RO). However, several other membrane desalination processes exist including Nanofiltration (NF), Ultrafiltration (UF), and Microfiltration (MF). Major thermal desalination processes include Multistage Flash (MSF) Desalination, Mechanical Vapor Compression (MVC), and Multiple Effect Desalination (MED). In this section, MSF will be discussed in detail, while the unique features of MVC and MED will be briefly summarized.

In a typical membrane desalination process, electric energy is used to drive the feed pump, thereby increasing the feed pressure. The feed pressure and membrane selectivity allow for the passage of fresh water through the membrane and the rejection of more than 98% of the dissolved salts in the feed. The rejected brine is then passed through an energy recovery unit to increase system efficiency.

Thermal desalination processes are mainly driven by heating steam; however, electric energy is used to operate various pumps and control systems. In a thermal desalination process, the heating steam is used to evaporate fresh water from the saline feed through either boiling or flashing.

In RO desalination, the saltwater feed is introduced at high pressure on the retentate side of the RO membrane while the permeate side of the membrane is maintained at much lower pressure. The pressure differential between the retentate and permeate sides drives reverse osmosis because it is greater than the osmotic pressure of the saltwater feed. The typical RO process consists of four main steps: (1) feed water intake, (2) feed pretreatment, (3) reverse osmosis desalination, and (4) posttreatment. In the first step, an intake pump pumps feed water from the source to the RO plant. Possible pretreatment steps include coagulation, chlorination, dechlorination, addition of acid for pH adjustment, addition of antiscalants, and filtration. The pretreated feed is then passed to RO membrane modules using a high-pressure pump. The reject brine from the retentate side is passed to an energy recovery device such as a pressure exchanger. Finally, the fresh water permeate is posttreated. This involves chemical addition for disinfection, adjustment of hardness and pH, and possibly filtration through lime.

In MSF desalination, the saltwater feed is distilled by flashing a portion of the feed water to vapor across several flashing stages. MSF desalination also consists of four main steps: (1) feed water intake, (2) feed pretreatment/and deaeration, (3) flashing stages, and (4) posttreatment. The saltwater feed is first pumped through a large duct containing course screens to remove large suspended solids. During pretreatment, the feed is chlorinated and treated with antiscalants and antifoaming agents. The feed is also deaerated to remove dissolved gases; the low thermal conductivity of these gases would otherwise reduce the rate of heat transfer around condenser tubes. Dissolved oxygen and carbon dioxide could also contribute to corrosion. It should be noted that flash stages are connected to vent lines through which any remaining non-condensable gases are passed to a vacuum steam ejector. The feed is the passed to a brine heater where it emerges as heated brine. The heated brine then enters a series of flashing stages in which a portion of the brine is flashed to vapor. In each stage, this small amount of water vapor condenses around condenser tubes, accumulates in the distillate tray, and flows to the next stage. This distillate stream also flashes off generating a small amount of vapor as it flows from stage to stage, further heading the feed seawater stream. In the final stage, the brine blow down is collected and rejected back to the water source while the distillate is posttreated through chlorination and adjustment of its pH.

Like MSF, MVC desalination is also a distillation process. However, in MVC, the heat used to evaporate the saltwater feed is provided by mechanically compressed vapor. The unique feature of MVC is that it is solely driven by electricity (which is used to power the mechanical compressor, as well as pumps, controllers, and the vacuum steam ejector).

MED desalination is another distillation process through which steam is circulated through a series of tubes (each tube is part of an 'effect'). In each effect, steam is condensed on the inside of the tube while a portion of the saltwater feed is evaporated on the outside.

7.4.4.2 Design Equations for System Sizing

Recall that the first step in selecting the appropriate HFCD system among possible technology types and integrations is to determine the (peak) electricity and water demands. This section will present the system design equations that will be used to size the system given this information. A typical HFCD system is diagrammed in Fig. 7.6. This type of system will serve as the basis for the design equations.

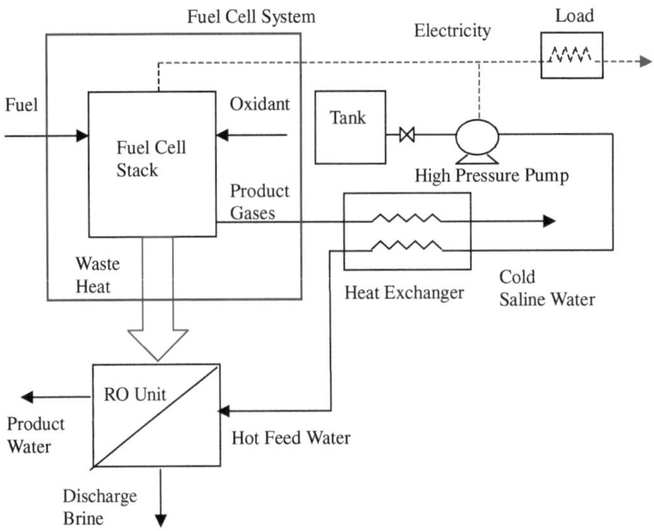

Fig. 7.6 Conceptual design for a typical hybrid fuel cell/desalination system [6]

The outputs from the fuel cells will be treated as simple variables in this exercise. The cooling air from each fuel cell will have a variable temperature, but the temperature will be an independent variable in the scope of the analysis and will not depend on fuel utilization, pressure, or any other factors. The difference between the fuel cell temperature and the desalination temperature will determine the power (Q) that can be exchanged between the cooling air and brine streams, as shown in Eq. 7.1.

$$Q = \dot{m}\, C_{\min}(T_{\mathrm{FC}} - T_{\mathrm{Desal}}) \tag{7.1}$$

where T_{FC} is the fuel cell temperature, T_{Desal} is the top brine temperature of the desalination process, $C_{\min}$ is the lower specific heat capacity between the water and the air, and $\dot{m}$ is the mass flow rate of the cooling air.

These two temperatures are dictated by what type of fuel cell and what type of desalination unit is used. Table 7.1 shows temperature ranges for various fuel cell and desalination unit types

The mass flow rate of cooling air is also dependent upon the type of fuel cell, as they have different ratios of electric power versus thermal power. The power ratios for 3 fuel cell types are presented in Table 7.2.

As an example of how the data from Tables 7.1 and 7.2 can be used, note that a 1 MW PAFC plant will provide 1.27 MW of thermal power, and would operate at temperatures between 80 and 200°C. Proceeding with these values, the mass flow rate can be obtained from Eq. 7.2.

$$Q = \dot{m}\, h(T) \tag{7.2}$$

where $h(T)$ is enthalpy as a function of temperature.

The 1 MW PAFC plant from our example would require approximately 11 kg/s of cooling air to dissipate the 1.27 MW of thermal power when operating at 90°C. This mass flow rate can then be plugged into Eq. 7.1 to obtain the maximum power that can be transferred to the brine.

The amount of heat that can be transferred to the brine is also dependent upon the temperature difference between the air and the brine. The most common

Table 7.1 Temperature ranges of fuel cells and desalination units [2]

System	Temperature (°C)
PAFC	80–200
MCFC	300–700
SOFC	700–1,000
MSF	110
MED	70
RO	4–40
MVC	–

Table 7.2 Power ratios [2]

	PAFC	MCFC	SOFC
Power ratio, heat:electric	1.27:1	1:1	0.67:1

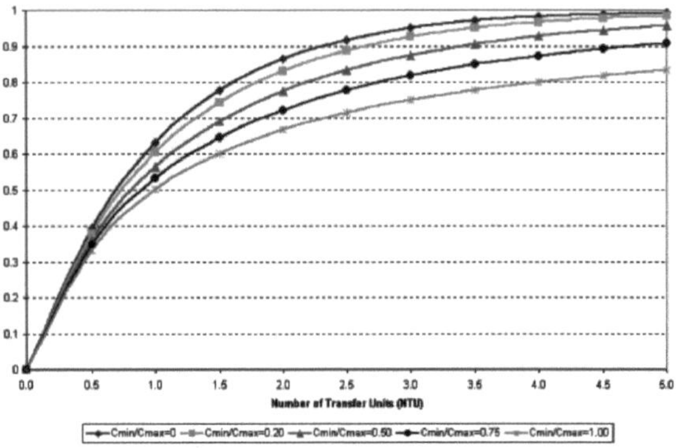

Fig. 7.7 Heat exchanger effectiveness

method of transfer heat from a hot gas to a cooler liquid is to use a shell-tube heat exchanger. Due to their modular nature and relatively small volume, it should be possible to locate them very close to where the intake water needs to be heated, allowing for minimal heat loss. Equations 7.3 and 7.4 show how much of this power can actually be transferred to the brine.

$$\varepsilon = \frac{Q}{Q_{\max}} \tag{7.3}$$

$$\varepsilon = \frac{1 - e^{-N(1-C)}}{1 - Ce^{-N(1-C)}} \tag{7.4}$$

where ε is the effectiveness of a heat exchanger, $N = UA/C_{\min}$ is the number of transfer units, and $C = C_{\min}/C_{\max}$.

In this case, C is known, but potential UA values have a large range, so the following chart can give an approximate effectiveness range. For a C value of 0.24, an effectiveness range of 0.6–0.9 is reasonable [7]. The results of Eq. 7.3 for several values of C are shown in Fig. 7.7. The optimal UA value will be determined by weighing the benefit of a more effective heat exchanger against the higher initial cost.

Furthermore, for each desalination process, the heat required to create every cubic meter of fresh water (specific heat) is known (see Table 7.3 below). When this value is divided by the available heat from air at a given temperature, the ratio of air mass flow rate to product water mass flow rate can be determined (see Eq. 7.5). The mass flow rate of water is dependent upon how much water is to be produced each day.

$$\frac{\dot{m}_{\text{water}}}{\dot{m}_{\text{air}}} = \frac{Q_{\text{air}}}{Q_{\text{water}}} \tag{7.5}$$

Table 7.3 gives values for the amount of heat and electric energy required to produce one cubic meter of desalinated water:

At this point, the necessary mass flow rate of heated air is known, so the overall size of the fuel cell system can be determined using Eq. 7.6. Note: the sizing process for RO is slightly different as the required system size based on thermal analysis will always be smaller than based on electrical analysis. Also, the heat and electricity values for RO are given as a range since the amount of heat required is a strong function of seawater intake temperature.

$$P = \dot{m}_{\text{air}}\, hR \tag{7.6}$$

where R is the ratio of heat to electricity for a given fuel cell system as given in Table 7.3.

This equation sizes the system such that adequate thermal power is provided for the desalination process. This means that there will be excess electrical power available that can be sold to the grid. This excess electrical power can be determined from Eq. 7.7:

$$P_{\text{excess}} = P - \frac{e_d V}{24} \tag{7.7}$$

where V is the volume of water produced per day and e_d is the specific electrical energy required to desalinate one cubic meter of water.

7.4.4.3 Results of Simulation and Economic Feasibility Study

This section presents the results of a simulation similar to the one outlined in the methodology (in addition to results from the ensuing economic feasibility study). The following tables (Tables 7.4, 7.5) show the results of a simulation run to determine the necessary fuel cell system size to produce 100,000 m^3/day of desalinated water from each of the four desalination methods. In order to provide 100,000 m^3/day, the desalination system needs to have a maximum capacity of 120,000 m^3/day.

Table 7.3 Energy demands for desalination [6]

	Electric (kWh/m^3)	Heat (kWh/m^3)	Heat (kJ/kg)
MSF	4	14	50.4
MED	4	11	39.6
RO	4–5	4–5	15.4–19

Table 7.4 Inputs

	Temperature (°C)
PAFC	160
MCFC	650
SOFC	900
Seawater	28
	Price
Natural gas	\$9/MMBtu
Electricity	\$0.13/kWh
Water	\$0.7/m^3

Table 7.5 Outputs

	PAFC	MCFC	SOFC
Fuel cell system size (MW)			
MSF	N/A	109.71	156.15
MED	100.25	80.26	116.77
RO	36.31	29.07	42.30
MVC	62.50	62.50	62.50
Excess electrical (MW)			
MSF	N/A	71.1	108.3
MED	63.5	47.5	76.8
RO	12.4	6.6	17.2
MVC	0	0	0
Fuel cell initial cost (MM $)			
MSF	N/A	307.2	546.5
MED	350.9	224.7	408.7
RO	127.1	81.4	148.1
MVC	218.8	68.8	218.8
Daily costs (thousands $)			
Desalination (not including energy) (thousands $)			
MSF	N/A	33.6	33.6
MED	33.6	33.6	33.6
RO	47	47	47
MVC	84	84	84
Fuel cell (O&M and fuel) (thousands $)			
MSF	N/A	209.8	284.8
MED	220.4	153.5	213.0
RO	798.3	55.6	77.2
MVC	137.4	119.5	114
Total operating cost (thousands $)			
MSF	N/A	243.4	318.4
MED	254.0	187.1	246.6
RO	126.9	102.6	124.2
MVC	221.4	203.5	198.0
Incomes (thousands $)			
Daily	70		
Daily electric income (thousands $)			
MSF	N/A	221.8	337.7
MED	198.2	148.3	239.5
RO	38.6	20.6	53.6
MVC	0	0	0
Daily total income (thousands $)			
MSF	N/A	291.8	407.7
MED	268.2	218.3	309.5
RO	108.6	90.6	123.6
MVC	70.0	70.0	70.0

(continued)

Table 7.5 (continued)

	PAFC	MCFC	SOFC
Daily net income (thousands $)			
MSF	N/A	48.4	89.3
MED	14.2	31.2	62.9
RO	−18.2	−12.1	−0.6
MVC	−151.4	−133.5	−128.0
Annual net income (MM $)			
MSF	N/A	17.7	32.6
MED	5.2	11.4	22.9
RO	−6.7	−4.4	−0.2
MVC	−55.3	−48.7	−46.7
Simple payback period (years)			
MSF	N/A	17.4	16.8
MED	67.5	19.7	17.8
RO	N/A	N/A	N/A
MVC	N/A	N/A	N/A

As suggested by the methodology, the outputs of the simulation have been used to calculate economic parameters for the project. Tools for economic analysis will be elaborated in Sect. 7.6. However, for now, it is important to know that payback period refers to the time in which the project will earn back the value of the original investment. For details on how to determine the various incomes, refer to Sect. 7.6.

It appears that both the RO desalination and the MCFC methods are at a large disadvantage in this study. The key problem with this method is that the value of selling electricity is valued much more highly than putting it into water. If electricity prices are at \$0.10/kWh and water prices are at \$0.6/m^3, then the 4 kWh of energy that it takes to make one cubic meter of water, could be sold for \$0.40, rather than being contributed towards the water production. The non-energy costs of desalination are greater than \$0.20/m^3, so it is unprofitable to use electricity for desalination when it could be sold.

The RO and MCFC systems are at disadvantage because they require such small systems to provide the necessary water production that there is very little excess electricity.

7.5 Proposed Solution and Justification

The proposed solution to Caye Caulker's water and energy problems is to replace the diesel generators currently used with a molten carbonate fuel cell/reverse osmosis (MCFC/RO) HFCD system.

Molten carbonate fuel cells are typically used for large-scale stationary power generation applications. The high efficiency of molten carbonate fuel cells, along with their ability to utilize heat for cogeneration makes them an ideal choice for this project. Molten carbonate fuel cells are also relatively well developed compared to other fuel cell types used for stationary power generation.

A key benefit of Reverse Osmosis is that, although it benefits from higher intake temperatures, it can be completely independent of the thermal waste of the fuel cells. Thus, unlike thermal based desalination methods (e.g. MSF), RO systems do not require a large degree of oversizing. Also, by making use of excess electricity, RO systems eliminate the need for expensive electricity storage systems. In an MSF system, if water is desperately needed, then the fuel cell will have to overproduce electricity to produce enough waste heat to drive MSF. The excess electricity will either have to be thrown away or stored. If an RO system is in this same situation, the fuel cell system can still increase its power output, but this power can go directly to the RO process. The additional energy will be stored in water, which is already a desired product. This unique ability may be the largest upside of using RO over any other method. Modularity has also been mentioned as a benefit.

It is assumed that the price of natural gas on an island is \$11/MMBtu, which is higher than most of the world. It is also assumed that the product water can be sold for \$0.50/m^3.

The feasibility of this solution is demonstrated in the following example. If a 600 kW MCFC system is installed on Caye Caulker, and the goal is to provide 600 m^3/day, then 100 kW of electricity and the excess from 140 kW heat will go towards water production. This leaves 500 kW of electricity and 460 kW of heat for the rest of the island. If the island power demand is 575 kW, then 225 kW of the excess heat can be converted into 75 kW of power and the other 235 kW can be devoted to other purposes, and will be available for cogeneration if needed. If less water is needed, then the system can be sized even smaller.

7.6 Economic Feasibility

This section introduces some concepts that will be necessary for economic analysis, and then applies some of these analyses to the 600 kW system proposed for Caye Caulker in Sect. 7.5.

7.6.1 Tools for Economic Analysis

Equation 7.8 introduces a very important parameter for combining system outputs such that profits are optimally maximized. This parameter (R) is the ratio of water cost to lost electricity cost.

$$R = \frac{W_c}{E_p\left(e_d + \frac{H_{\text{excess}}}{C}\right)} \tag{7.8}$$

Table 7.6 Fuel cell costs [8, 9]

	PEM	PAFC	MCFC	SOFC
$/kW Installed	5,000	4,500	2,800	3,500
O&M ($/kWh)	0.023	0.029	0.033	0.023
Fuel ($/kWh)	0.4	0.0475	0.037	0.04

Table 7.7 Desalination costs ($/m^3)

	MSF	MVC	MED	RO
O&M	0.2	0.2	0.2	0.28
Energy	0.3	0.3	0.3	0.22

where E_p is the local electricity price, H_{excess} is the amount of excess heat in kWh, and C is the conversion factor for how much of the excess heat could be converted to electricity in a cogeneration unit.

This factor will vary with temperature, but current cogeneration units in operation are able to convert about 1/3 of the heat to electricity. If R is greater than one, more water should be produced. If R is less than one, the excess heat should be used to generate electricity that will be sold to the grid.

The daily income of the system is the sum of the incomes from both electricity and water production. The net daily income is the total income minus the daily operating costs. For fuel cells, the daily operating costs can be broken down into operating and maintenance costs and fuel costs. Table 7.6 shows how these costs compare for each fuel cell type along with the initial capital cost.

Table 7.7 similarly shows the cost breakdowns for desalination systems. It is assumed that the cost of each desalination process can reach up to $0.5/m^3 in total. The percentage of the total cost coming from operations and maintenance costs versus from energy costs vary with each individual system; the values in Table 7.4 are averages.

These costs are the production costs over the projected lives of the units. Thus, the initial construction costs are imbedded in the operating and maintenance costs. The energy cost for desalination will be completely eliminated by the fuel cell system. The price of desalinated water is irrelevant to this analysis since water is largely subsidized by the government.

To determine the daily income of the system, the two portions of the desalination cost (O&M,Energy) must be treated as either fixed values or fixed ratios. That is, if the desalination cost of an MSF plant is $0.7/m^3, will the energy cost still constitute 60% of the unit cost, or will it still be $0.3/m^3? The economic analysis results tabulated in Sect. 7.4.4.2, they are treated as a fixed ratio. Adding the desalination costs to the fuel cell costs gives the total daily cost. To calculate annual income, the net daily income is multiplied by 365.

The other additional cost that will be considered is the cost of a gas turbine for cogeneration. Gas turbines can be installed in the range of $1,500/kW. A range of $1,000–$2,000/kW was used for the analysis in Sect. 7.4.4.2.

Equations 7.9–7.11 provide tools to help analyze the economics of the Caye Caulker project.

$$PV = A\left[\frac{1 - \frac{1}{(1+i)^n}}{i}\right] \tag{7.9}$$

PV is the present value of an annuity, *A*, that is compounded at an annual interest rate of *i* for *n* years. In this case, the annual income will be the value of the annuity. The interest rate is a user-defined input based on their current economic situation. The length of the project is the system life of either the fuel cell or the desalination system, depending on which is shorter.

The *PV* value can be plugged into Eq. 7.10 to yield the return on investment.

$$ROI = \left(\frac{PV}{C} - 1\right) \times 100 \tag{7.10}$$

ROI is the return on investment, *C* is the original capital cost of the investment. This number gives the percent of return on the initial investment. An *ROI* of 15% means that for every dollar invested in a project, the investor receives $1.15 back.

$$PP = \frac{C}{A} \tag{7.11}$$

PP is the payback period, or the amount of time it will take for the project to earn back the original investment. The payback period for traditional power plants is in the three-year range. As a point of comparison, solar arrays typically have payback periods of 5–10 years.

7.6.2 Economic Feasibility of Caye Caulker HFCD

The 600 kW MCFC system proposed in Sect. 7.5 would cost about $2 million and would provide an annual income of $400,000. With this income, the system would pay for itself in around four and a half years. The graph in Fig. 7.8 plots payback period as a function of the electricity price and natural gas cost. The graph shows that the system economics are very sensitive to the price at which the electricity can be sold. This analysis assumed that the desalinated water could be sold at $0.70/m^3.

As shown in Fig. 7.8, the payback period can extend to four times the minimum if electricity prices are low. When the price of electricity goes above $0.15/kWh, the payback period is reduced to around 5 years. Although this electricity price is high for developed areas, it is low for remote areas that rely solely upon diesel generators.

Figure 7.9 plots payback period against water cost. The payback period is not nearly as sensitive to water cost as it is to electricity price. The payback periods only vary by 15–50% even when water cost doubles. However, similar to the

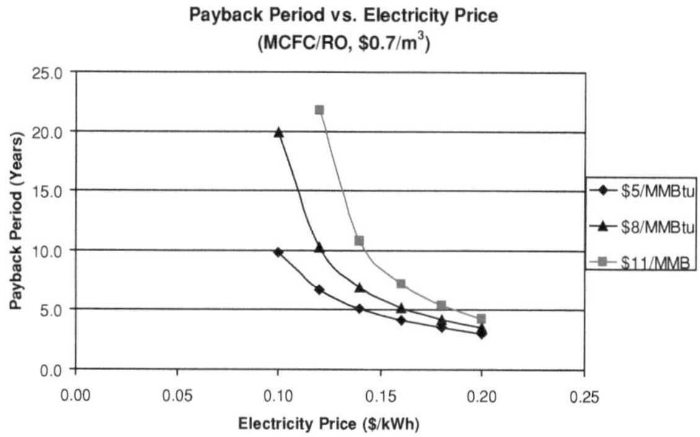

Fig. 7.8 Payback period as a function of electricity price and natural gas cost

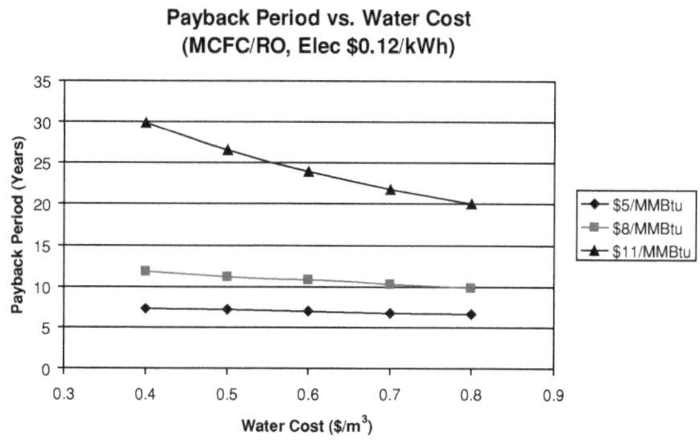

Fig. 7.9 Payback period as a function of water cost

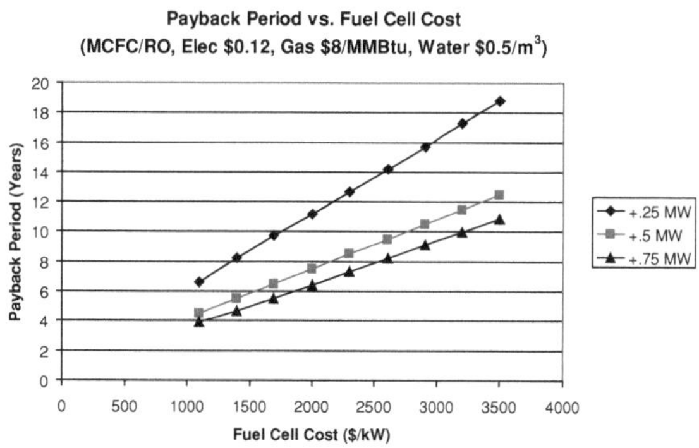

Fig. 7.10 Payback period as a function of initial fuel cell cost

results shown in Fig. 7.8, the payback period is more sensitive to natural gas cost especially at low electricity prices.

Finally, Fig. 7.10 shows how sensitive the payback period is to initial cost of the fuel cell. Each of the three data series in the graph represents a different degree of oversizing from the size needed to meet the water requirements. This plot assumes natural gas costs of $8/MMBtu, and electricity prices of $0.12/kW.

Taken together, these three plots illustrate that the economics of the HCFD system are very dependent on both the amount of electricity that can be sold, and the initial cost of the fuel cell. The project will be economically feasible at this relatively low electricity price if the installed fuel cell cost can be as low as $1800/kW, which is not far from the fuel cell costs that are forecasted over the next 5–10 years.

7.7 Concluding Remarks

In detailing the process of finding a solution for Cayce Caulker's problems of water scarcity and insufficient electrical power, this case study demonstrated that it is reasonable to use waste heat from fuel cells to aid desalination. Hybrid Fuel Cell/Desalination (HFCD) systems are an appropriate and attractive means of supplying developing regions with adequate electrical power and water. HFCD technology is best used in remote locations where the modularity and small scale efficiency of fuel cells is most optimal.

The methods applied to the Caye Caulker case study can yield similar (or even better) results when applied to other regions. For an area like Eritrea, the natural gas prices would be even lower, which will further improve the economics. Also, Failaka Island, the Arab Gulf Region, Rhodes, Greece, and many Mediterranean islands have conditions similar to Caye Caulker, but provide easier access to natural gas.

The chapter also introduced a step-by-step methodology for sizing, selecting, integrating, and optimizing an HFCD system. This methodology allows one to choose an appropriate HFCD system for any number of applications, provided that economics are always considered as the primary decision making criteria.

References

1. Region faces up to the urgent need for increased water production and strengthened infrastructure. AME Info: The Ultimate Middle East Business Resource, 18 June 2006. http://www.ameinfo.com/89157.html. Accessed 2 April 2007
2. EG&G Services, Parsons, Inc, and Science Applications International Corporation (2000) Fuel Cell Handbook, 5th edn. DOE/FE/NETL, Morgantown, WV
3. Al-Hallaj S, Alasfour F, Parekh S, Shabab Amiruddin J, Selman R, Ghezel-Ayagh H (2004) Conceptual design of a novel hybrid fuel cell/desalination system. Desalination 164:19–31
4. Ali M, El-Nashar (2000) Role of desalination in water management in the gulf region. Abu Dhabi Water and Electricity Authority, Abu Dhabi
5. "Electricity Demand-side Management" New Zealand Treasury, 6 March 2006. http://www.treasury.govt.nz/electricity/edm/7.asp. Accessed April 2007
6. Al-Hallaj S, Rob Selman J (2007) A study of hybrid fuel cell/desalination systems. Final report MEDRC-03-AS-007, Middle East Desalination Research Center, May 2007
7. "Remembering How and When to Use Heat Exchanger Effectiveness" Chemical Engineers' Resource Page, 2004. http://www.cheresources.com/hteffzz.shtml. Accessed 2 April 2007

8. Houston Advanced Research Center Fuel Cell Industry Assessment. 2001. http://files.harc.edu/Projects/FuelCell/Reports/IAR/2001FuelCellIndustryAssessment.pdf. Accessed April 2007
9. Monis Shipley A, Neal Elliott R (2007) Stationary fuel cells: future promise, current hype. American Council Energy Efficient for an Energy-Efficient Economy, March 2004. http://www.aceee.org/pubs/ie041full.pdf. Accessed April 2007

Index

MIX
Papier aus verantwortungsvollen Quellen
Paper from responsible sources
FSC® C105338
FSC
www.fsc.org

If you have any concerns about our products,
you can contact us on
ProductSafety@springernature.com

In case Publisher is established outside the EU,
the EU authorized representative is:
Springer Nature Customer Service Center GmbH
Europaplatz 3, 69115 Heidelberg, Germany

Printed by Libri Plureos GmbH
in Hamburg, Germany